BoatSense

BoatSense

Lessons and Yarns from a Marine Writer's Life Afloat

DOUG LOGAN

BoatSense

Lessons and Yarns from a Marine Writer's Life Afloat

Seapoint Books + Media LLC

PO Box 21, Brooklin, ME 04616

www.seapointbooks.com

Design and composition by Claire MacMaster

barefoot art graphic design

Printed by Printworks Global Ltd., London & Hong Kong

First Edition

ISBN: 978-1-7325470-1-8

For Melissa, Nick, and Jane

The tide is ebbing in the river. The current curls past the docks,
carrying froth green with pollen. Over the marsh willets rush
and shrill, and fiddler crabs scuttle the flats. The osprey is back
on her perch, returned for spring from the Spanish Main.
Let's idle downstream together. Never a finer day.

— DOL

LIST OF TECHNICAL SIDEBARS

Introduction

This collection of yarns, prescriptions, lists, avowals, and explorations is for people who love boats, particularly sailboats, and the skills and sensibilities that a life around boats can help develop—things like self-reliance, adaptability, a sense of balance between daring and humility, and an appreciation of our place in the grand scheme.

The title, BoatSense, first has to do with the simple physical familiarity with how boats float and move in the water; how wind and waves act on them; how their parts work; how they're guided and managed. As kids who grow up on the water know, the best way to get the feel for a boat is to rock it. Just grab it and make it move. Stand on the gunwale, push down on the bow, pull on the shrouds. See how much effort and weight it takes to get things moving, and then notice how it moves. You can even do it with big, heavy boats if you try hard enough and get some momentum going. After you've put a boat through the rock-and-roll test, you'll have a better notion of how to get it in and out of a slip, or coast up to a mooring.

Second, boatsense involves the learned and ingrained habits of seamanship, which migrate successfully to dry land. On shore today there's a service industry devoted to coaching people how to get rid of their clutter and simplify their lives, but people who spend much time on boats already know what's necessary, how to pack it, where to stow it, and what to leave behind. On shore, the conservation movement strives to get people to understand what boat

> If your idea of a comfortable boat is one with air-conditioning, a washer-dryer, satellite TV, and a generator that needs to run for hours a day, we're of very different mindsets. You might think of me as some sort of hair-shirted masochist, and I might think of you as... exactly the same thing. Clearly, we solicit our pain from different ends of the spectrum.

people have always known: fresh water and food are finite. Fuel and electricity are finite. Care must be taken. Good habits must be developed.

If you're young and get bitten by the boat bug, you'll probably carry it for life. It can happen later too, of course, and sometimes then the condition can be more acute, more like an obsession. When you have the bug, you think a lot about being on the water, what it's like there, what to have there, how to do things there. And when you're there you find, as did the Water Rat in *The Wind in the Willows*, that there's nothing half so much worth doing. To people with boatsense, boats are living things with traits and manners, strengths and weaknesses. Thinking about boats can turn into a meditation that calms you for sleep, and boats can be a refuge in your dreams.

There are plenty of people who know more than I do about pretty much everything, including boats. But I have learned over the years that the more vehemently a person claims expertise in a certain area, the more likely he is to be vehemently opposed by a competitor with equal claims. This causes confusion. I have also learned that boats, their systems, and the waters they travel upon are humbling in the long run, and that people with the most useful experience and understanding are the least insistent and abrasive. I hope that I am one of those, and therefore would like to disavow expertise of any kind.

Instead, my experience has been more that of a long-term addict who gradually accumulates a bit of wisdom based on hard knocks. I grew up on the water and in boats, and worked summers in a boatyard, swabbing the heads, moving blocking and cribbing, helping to haul and launch, and then moving into more responsible work. I learned the rudiments of painting, varnishing, and fiberglassing. I stepped and unstepped masts, set up standing and running rigging, learned to splice laid rope and braid. I learned something about tools and fasteners. I learned to move boats under their own power and under tow.

In my 20s I lived aboard a much-used 33-foot yawl (later a sloop, later still an engineless sloop) at City Island in the Bronx. I learned a great deal about parts and systems aboard that boat, often by breaking them first. Some things I fixed; others I could do without—including, eventually, the engine.

My work as a boating writer and editor has allowed me many years of living the Water Rat's dream under sail and under power. I've had a chance to race and cruise inshore and offshore, and gotten to test all sorts of boats and boat gear. Despite all that, and all the will in the world, I'm no master shipwright. I'm always pleased when an electrical repair or installation of mine actually works. I avoid fiberglassing if I can, because I usually glue myself together. I have neither the chops nor the tools to do much with wood if it isn't a piece of plywood or a two-by-four. However, over the years the hard knocks have done their trick, and I know my own bad habits well enough to be able to mitigate and even prevent some of my DIY disasters.

Partly because my own skills are not profound, but even more because I've seen plenty of hard knocks that were not my own, I've developed some strong views that underlie the idea of boatsense, and here any apologia stops.

I don't believe that a boat should try to be a floating imitation of a house ashore, or a hotel ashore, or a car ashore. I am against, for several reasons, bringing power-hungry, complicated, land-based "comfort" systems onto a boat. Instead, I favor simplicity, even a certain degree of hardship and inconvenience, to a clutter of systems and gizmos. In profusion, they inevitably divert too much attention from what, to me, are the essential attractions of being afloat—the boat's relationship to wind and water, the surroundings, conversation with friends, and the habits of seamanship and self-reliance that help a person live happily both afloat and ashore.

If your idea of a comfortable boat is one with air-conditioning, a washer-dryer, satellite TV, and a generator tha needs to run for hours a day, we're of very different mindsets. You might think of me as some sort of hair-shirted masochist, and I might think of you as...exactly the same thing. Clearly, we solicit our pain from different ends of the spectrum.

Pursued in a way that will give lasting pleasure and benefit, this isn't a game for everyone. Years ago, I wrote an opinion piece for a boating industry publication saying that we in the industry would do better for our customers and ourselves in the long run by not always hyping boats with photos of bikini-clad beauties lounging on the foredeck in calm, gin-clear water, and instead by playing up the idea that this was more often a challenging, think-on-your-feet pastime, with a lot of learning involved. I was promptly called to task by the marketing brigade, and clearly their point of view has held sway because boating ads are pretty much the same as they've always been. But I stand by what I said: Boats aren't for everyone, and those who are lured into boats purely by promises of ease and comfort will eventually be disappointed.

But those who persevere through the hard parts learn something about themselves that mere landsmen can never know. Someone said that a day spent sailing is so good that it doesn't even count in your earthly total, as if Lachesis, the Fate who measures the thread of life, takes a breather and looks the other way. (The fact that "good" can mean "good for you," not necessarily pleasant, is understood.)

Most of this book is about sailing and sailboats, but if you think all the time of paddling your kayak or rowing your ocean shell or fishing out of your Grady White, you've got the bug. We speak the same language.

While there are specific bits of advice here, this book is in no way intended to cover the range or depth of information in the big, standard boating references available. There's a short list of can't-do-without practical books in the back.

A couple of usage notes. I use the word "rope" when speaking at the generic level. A stickler would say even then that the word refers to cordage over an inch in diameter. Although there are some specifically named rope bits on boats, like a bolt-rope, generally when a length of rope has a purpose on a boat, it becomes a "line," and line is further broken down into halyards, sheets, anchor rode, and so on. For cordage under about a quarter inch, like flag halyard, parachute cord, whipping thread, and tarred marline, I like the old-fashioned term "small stuff."

I write "dead reckoning" instead of "ded reckoning," even though the term is derived from "deduced reckoning."

Gender roles on boats can be an issue, and they can be an issue in writing about boats. With all respect for the motives behind attempts to unisex-ify words like "helmsman" I'm going to stick with the traditional. Same for "man overboard." Some of the best sailors in the world are, have been, and will be women, and most of the worst are men. I've sailed as crew for women who are among the finest sailors and seamen, and without exception they have used the terms "man-overboard" and "helmsman."

Acknowledgments

Most of what follows here originally appeared, in somewhat different form, in *Practical Sailor* magazine, *Sailing World* magazine, and the websites of the Boats Group. Many thanks to my friends at those organizations for publishing the material originally.

CHAPTER 1

People, Places, and Yarns

What draws people to boats? Initially some get involved out of a sense of adventure—the taking of risks and the chance to experience things out of the ordinary. Some are fascinated by the how and why of marine shapes—the hulls, propellers, sails, keels, and rudders that operate in the water and the wind. Some love the camaraderie found among a boat's crew. Some like to fish. Some like to ski or wakesurf or wakeboard or tube. Some like to find a quiet place to drop the hook and chill out. Eventually, though, if people stick with boats, they get to enjoy all those things and more.

Here's a theory of existence: When you go Aloft, they check you for stories. They want to know what you've tried, noticed, and learned on the old planet. They find out whether you've whiled away your years in a lounger with a game console in your hands or challenged yourself in the actual world instead. You get credit not just for achievement, but for effort, and for the degree of difficulty of things attempted, and for knowing the difference between adventures (which are calculated risks) and hare-brained schemes.

You notice, as you're standing there waiting for the ethereal being with the clipboard, that the old saying was absolutely correct—you didn't bring a single thing with you. Your car is behind you. Your clothes and your dishes are behind you. Your new widescreen TV, your smartphone, and even your beloved boat are back there. All those things that seemed important are not with you. All you have is the stories you can tell. In your case and mine these will include some sea stories. If we're lucky, our interlocutor will be a sailor, too: "Well, now, bucko, spin us a yarn of your voyage."

As a rule, sailing adventurers are a joy to be with, because they are repositories of such good stories. I don't think I've ever met a devoted sailor, offshore or coastwise, who was fearful of taking on a well-considered challenge on land or sea, lazy about handling it, or particularly bashful about sharing the yarn later. Still, we all have our limits. I, for one, am not interested in sailing too near ice unless it's surrounded by bourbon in a mug.

I had a friend who was an offshore sailing instructor. He would take people, as you would suspect, out to sea. They gulped, as everyone does, when the land dropped under the horizon and they found themselves in a very big place on a small boat. They gulped again when the sun set on their first day at sea.

They kept heading offshore. They settled into the watch rotation. My friend would turn off all the electronics for long periods of time, making everyone keep a dead-reckoning plot. This settled some people down, but it increased the gulping sensation in others, who never got over that feeling of near-panic when they were out there. My friend said that maybe half the people who headed offshore decided (some very quickly) that being offshore was not for them. But that was OK, because at least they had found out what it's like. They'd seen it first-hand. They'd had an adventure, and now they had a story that would stay with them forever.

There are many levels of intensity in sea stories, ranging from the hilarious to the horrifying, but really, at the heart of most sea stories are the elements of difficulty and risk. Whatever you're doing, whether it's rewiring a bilge pump in 100-degree heat or changing headsails in mid-ocean while trying not to be swept off the deck, you're doing things that will always stick in your mind because they're apart from the ordinary. And who knows—these may be the stories that will get you into some sort of Advanced Program.

Pat and the Rat

I didn't have much trouble with rats when I lived aboard at City Island, the Bronx, back in the early '80s. I'd see them once in a while in the boatyard when I came home late on a summer night, hardly ever in the winter. A rat faced me down once when I was on the gangway headed to the dock. He was already on his way up. We both stopped. I took another step forward, then he took another step forward. He was a big rat, as well-nourished dock rats tend to be, and after a short standoff I backed away and let him come up. He was pretty stately about it, too.

The last time I saw a rat in that boatyard—the place was called Thwaites, but is now gone—he dropped through the foredeck hatch and landed on my bunk. It was a hot summer night and I was in the main cabin, naked, reading a book. I heard him land on the cushions and saw the glow of his eyes, and I was unhappy because I figured the only way out for him was across my lap and up the companionway. I scrambled for the fire extinguisher—the only thing within reach that I could use as a weapon—but when I turned back the glowing eyes were gone. There weren't too many places for a rat to hide on my boat, but I looked thoroughly because of what had happened to Pat a month or so earlier that summer.

Pat lived on an ancient sailboat about 36 feet long—spavined, hogged, and fire-engine red except where streaked with brown rust. It was, in my view, the flagship among the boats we lived aboard at Thwaites, some of which were eyesores, some good-looking, most just regular designs on their way either to refurbishment or decrepitude. But as both a home and an evolving project, Pat's boat floated alone. He was a good marine carpenter who had bitten off a lot to chew.

One evening Pat came back to his boat and went to make himself a sandwich, and there was no bread. This was puzzling because Pat lived on sandwiches and he was sure he'd had a full loaf in the food locker. But he made himself something else for supper, and after that decided to do some fix-it work. He needed a tool he didn't use often, and went to find it underneath the forepeak berth. He moved the cushion, lifted up the locker cover, shined a flashlight in—and there was his bread, somewhat torn to pieces, in the company of a very large rat who looked up at Pat angrily and made a sort of blurred, scurrying movement in the direction of Pat's hand.

Pat, though not timorous by nature, dropped the locker cover and went to look for something threatening to put between himself and the rat. Having armed himself with a dustpan he went back to do battle, but when he lifted the cover again the rat was gone—wriggled through a hole in the bottom of the locker that normally surrounded a pipe, but with the pipe removed during boat repairs it had become the proverbial rat hole. So the rat had easy access to the bilge, and therefore to the engine compartment, the gear and sail lockers, the lazarette, and anywhere there was a chink or a limber hole big enough for him to squeeze through. Pat plugged every conceivable pathway into the boat's living spaces, and at the same time left the rat a wide choice of egress. Still, as he climbed nervously into his bunk that night he knew that the rat hadn't left, that the rat had the run of the boat and wasn't about to leave; and that he was going to have to face the animal again, *mano a raton*.

Pat slept with his sleeping bag zipped up around his head, as most of us all did in wintertime. Meanwhile we were all glued to our seats. Some were full of advice: Poison it, snare it, lure it, deafen it with high-pitched noise, smoke it out, shoot it (in a boat, no less). But for Pat this was a point of honor: He would trap the rat alive. It was a nice idea. He made a simple have-

> He was a big rat, as well-nourished dock rats tend to be, and after a short standoff I backed away and let him come up. He was pretty stately about it, too.

a-heart trap—a wooden crate propped up by a stick with bait tied to it. When he checked it the next morning he found that the rat had delicately eaten the food off the stick where it stood. The next night he set the trap again. This time the rat sprang it. Pat was on the dock, and by the time he got down below the rat had jostled the crate over to the edge of the galley counter, run it into the fiddle there, toppled it over the edge, and gotten away. It was apparently a strong rat.

Pat went into the city the next day, leaving another trap, this one more complicated: The rat would have to fetch the bait on a false walkway that would collapse and drop him into a sailbag, whose neck would be cinched tight by his falling weight. That part worked. But by the time Pat got home the rat had evidently wearied of life in the bag, chewed his way out, and disappeared again.

A siege began. It lasted for quite a while as spring turned to summer and the docks began to swelter just north of the Keansburg, a mothballed excursion steamer that served nobly as a breakwater in the winter, but also as a windbreak when the fickle southwest zephyrs filled in from Eastchester Bay. Pat varied his trapping techniques, and the rat either defeated them or ignored them long enough to provoke Pat to change them. He made traps that would have made Rube Goldberg wince.

One day when I got home from work it was all over. The rat, as big as a cat, lay dead on the dock, his head stove in. Pat was sitting in the cockpit of his boat with a beer, receiving congratulations from the neighbors.

"Finally made a trap that outfoxed him! Good job, Pat!"

Pat just shook his head. "It didn't work out that way," he said.

What had happened was that Pat had used another wooden crate—and after he heard it fall off its propping stick he quickly ran over to it and slid another board under it. Then he picked the whole thing up and carried it up the companionway to the cockpit. He was going to carry it ashore and release the rat somewhere up in the boatyard, but his curiosity got the better of him and his lifted the crate off the bottom board just a bit to make sure the rat was in there. He saw no rat feet. He gingerly lifted each side of the box, a bit higher and a bit higher each time, thinking the rat was always moving to the opposite side. But it became apparent after a while that the rat was not in the crate at all and had made good yet another escape.

Pat then went to up-end the crate entirely, whereupon the rat, which had been clinging upside-down to the bottom, then let go, landed with a thud on the backing board, and took off like a rocket, aiming to get back down the companionway. But Pat sat, stunned, in its path. The rat hesitated, Pat's hand found the handle of the iron Hibachi grill that was sitting on top of the lazarette hatch, and flung it at the rat. It connected. The rat was stunned, then angry. The Hibachi grill clattered to the cockpit sole. Pat picked it up and threw it again, and this time it caught he rat squarely and knocked him against the aft end of the cabin.

Pat did not go into much further detail except to say that the battle had not ended there. The cornered rat behaved… well, like a cornered rat to the last. It was a short but horrifying fight, and Pat's cockpit had the dings and gouges and bloodstains to attest for it.

When I found Pat sitting with his beer, and the dead rat on the dock, I thought it was a bit gruesome. But after hearing the story I understood. As has been known since ancient times, a worthy enemy means a greater victory. You don't just kick such a crafty and elusive enemy over the edge of the dock; you lay him out— on a shroud embroidered by vestal virgins if you happen to have such a shroud—and contemplate him for a while.

So when the rat dropped through the foredeck hatch on my own boat, not very long after Pat's rat's demise, I did take a good look around, but then I decided to do nothing about it. If the rat stayed, he would be tolerated. I didn't want to see him, but if I had seen him I would have treated him pretty much as I'd treated the big rat on the gangway: "You first, mate. Here's a box of Ritz crackers. I'll just be heading up to the forepeak for the off-watch. Let me know if the wind changes."

I lived aboard for another year and a half. Sometimes I would think food was missing, and I'd blame the rat, although it was probably the rum. Six years after that I sold the boat to a nice guy from Florida, who had been looking for that particular hull—an uncommon semi-production design that could be fitted out any way you pleased from the cabin sole up. He told me he intended to remove the house and deck and rebuild from there. Ever since then I've imagined him with a reciprocating saw, buzzing slowly along the toe-rail, then turning inboard and opening up a cut at the foredeck hatch, only to discover a pair of close-set eyes staring up, not malevolent eyes, but authoritative and unblinking.

Fluids, Foils, and Flow

There is a poetry of sailing as old as the world.
— Antoine de Sainte-Exupery

Among the dozen or so topics in sailing that are worth a bit of serious study, from navigation and meteorology to diesel mechanics and electrical trouble-shooting, at the heart of the matter are aerodynamics and hydrodynamics, often linked under the common name of fluid dynamics. For sailors, this is the science of how the hull and appendages move through the water, and how the sails move through the air. Then comes the art and skill of doing what we can to make things more efficient.

Many years ago when I was working in New York I edited a book called *The Small World of Long-Distance Sailors*. It was written by a charming, soft-spoken woman named Ann Carl, who had recently returned from voyaging with her husband, Bill. She had much to say about the meetings they'd had in faraway harbors with people who had become detached from the land, some for long stretches, some forever.

The book was mostly a memoir, with some practical matters and a few observations about sailing and life in general thrown in. There were fanciful and beguiling illustrations by the author, and color photos by the author, and poems by the author. All these extra elements were good enough to accompany Ann's excellent writing.

It wasn't until the book was almost done, and I was sitting down to write the flap copy, that Ann, in an effort to provide some grist about herself, mentioned that she had been a WASP in World War II, and a military test pilot, and later the first woman to fly a fighter jet. Good grief, what a talented person. Later in life she became a strong advocate for conservation and the environment.

Over the years I've run across plenty of sailor-aviators. Sally Pinchot was a top-flight navigator as well as a pilot and musician. Jim Marshall, formerly of Ockam instruments, taught racing sailors the science of performance through instrumentation back in the '80s. Arvel Gentry, fabled Boeing engineer and Ranger 23 sailor, upended the common wisdom about how sails work back in the '70s. Steve Fossett, balloonist. Steve Dashew, glider pilot. Sir Francis Chichester and his *Gypsy Moth* airplanes and boats.

I think it's the fascination with foils and fluids. These are probably all people who spent more than their share of time as kids holding their palms flat outside the car window and flying them up and down, and noticing how different shapes make different wakes when you drag them through the water. It also has to do with navigation: Aviators and mariners are bound together

I think it's the fascination with foils and fluids. These are probably all people who spent more than their share of time as kids holding their palms flat outside the car window and flying them up and down...

by knots and nautical miles, 24-hour timekeeping, similar charts, and similar instruments. They view the earth, the atmosphere surrounding it, and the celestial bodies above it, in many of the same ways. They care deeply about weather, and about others traveling in the same medium. They are takers of calculated risks. They enjoy being skilled at endeavors that take years of practice, and that are not always comfortable.

At the highest levels, the fluid dynamics behind jet fighters and Formula One cars and America's Cup wing sails is the stuff of hard math and physics, chaos science, and computational horsepower. But for mere mortals who hoist their fabric wings to the breeze, a bit of fireside reading in the winter about things like lift, laminar flow, and angles of attack will take some of the mystery out of sail trim when summer comes. The next time you tighten a cunningham to move the mainsail draft forward, or move a genoa lead outboard to open the slot, or cinch down on the vang until the leech telltales are streaming right, you'll be thinking like a Wright brother—or a seagull.

Cuttyhunk Lightning Dance

Maybe I should have taken the shark as an omen. I saw its big dorsal fin circling in a patch of weed and trash in the chop south of Nashawena and Cuttyhunk.

Then there were the waves below Sow and Pigs. They weren't like Southern Ocean monsters, but neither was I riding in a Southern Ocean windjammer.

> I did a sort of St. Vitus' macarena, squatting and ducking behind the dash, daintily holding only the wooden parts of the wheel spokes with thumb and forefinger as if sipping tea...

Mine was a 26-foot diesel powerboat, climbing over steep, smooth, no-kidding 10-foot-plus hills, rolling in from a seemingly placid Atlantic on a standard summer afternoon, with the beginning of a fair tide coming into Vineyard Sound and a light southwest breeze on top. Weird. Just weird. These waves weren't close to breaking, I was glad to note, but I was interested in getting well clear of Sow and Pigs and making my turn into Buzzards Bay.

I knew better than to start a passage in mid-afternoon on a hot summer day in New England. Normally I wouldn't have. But you understand how it is. Circumstances prevent your departure. The bright lights of Newport on a hot summer night are beckoning. You lose your Yankee caution, rationalize, invoke the gods of luck and forbearance. But as we know, those gods only help those too ignorant to ask for their help. The rest of us have to be smarter.

The big waves settled into four-footers, which I took uncomfortably but, all things considered, gratefully on the beam, and things were looking up until I got to the midpoint of my 20-mile leg between Vineyard Sound and Brenton Reef, marking the entrance to the East Passage up to Newport and Narragansett Bay. Then the sky to the north abruptly turned gunmetal gray, and a high, curved stormfront extended rapidly to the south—a front that looked like it should be carrying the flying monkeys of the Wicked Witch. Then lightning began to connect the front with the Rhode Island and Massachusetts shoreline.

After a phase of denial during which I changed course and speed a few times to see if I could somehow avoid the sucker, I realized I wasn't going to. It was moving faster than I could move in any direction. The only way out was through. I looked all around to make sure no one else was damn fool enough to be around, made sure of my compass course, throttled back a bit, took a deep breath, and went on in.

The first few minutes were extra exciting. The initial blast of wind seemed to be about 40 knots and mashed the waves right down. It was dark except for the white froth on the surface of the flattened chop. Visibility was maybe 50 yards. Rain was making its way through the window gaskets. It was loud. There were three or four lightning strikes close by, and then the doozy—smell of ozone, hiss and tingle of static, hair going weightless, and the bolt that seemed to come horizontally over the boat from ahead. It hit somewhere behind, maybe 50 or 500 yards, I didn't see.

Years ago, Giff Pinchot told me he'd been at the helm when his boat was hit by lightning, and the flash had blinded him for several hours. This story had stuck with me, so for the whole squall I did a sort of St. Vitus' macarena, squatting and ducking behind the dash, daintily holding only the wooden parts of the wheel spokes with thumb and forefinger as if sipping tea, closing one eye and then the other to keep one eye in reserve, and waiting until after a flash to jab my hand up and twist the metal handle of the manual windshield wiper.

Not long after I'd polished my dance routine, the squall let up. The rest of the trip into Newport was wet and bouncy, but eventually a horizon appeared and there was a hint of late-afternoon blue to the north. The last half-leg turned out to be one of the most enjoyable times I've spent on the water. It's funny how life can seem rosy after the worst of the bad stuff has gone away and you're left with only the standard discomforts. Maybe I wasn't as glad as Captain Bligh was when he reached Kupang after his 3600-mile ride with 18 others in a 23-foot open boat, or as happy as Shackleton after his 800-mile Southern Ocean ordeal with five others in a 22-foot whaleboat. But Newport looked good to me that night.

There's No Place Like Home Waters

In the immortal words of Buckaroo Banzai, "Wherever you go, there you are." Maybe that's a simple reminder to have a good look around and pay attention, even if you're somewhere you've been a hundred times before.

To me, any place big enough to hold a few boats is an interesting one, whether it's the Sea of Cortez or Smith Mountain Lake in Virginia, or the Tennessee River, or Galveston Bay, the Chesapeake, or the Gulf Stream offshore. Every single patch of water, large or small, salt or fresh, has a history; every patch has drawn settlers, pilgrims, birds, and animals to its shores and watery creatures to its depths, from frogs and perch and water moccasins to seals and manta rays. So it's always good to be on a boat in an unfamiliar place and learn something about what makes it tick. But if seeing new things is worthwhile, so is taking a closer look at familiar sights.

My home base has always been Long Island Sound, a body of water so full of cool features and history that I could spend several lifetimes just grokking on everything. I've been up, down, and across it many times, racing and cruising, and the more I see of it, the more I appreciate it, and the more protective I feel about it. Here's a quick portrait.

It's about 100 nautical miles long by 20 across at its widest point, with an estimated 18 trillion gallons of water, much of it flooding and emptying twice a day, and tides ranging from a little over two feet to over seven. Three major freshwater rivers converge with it, including the Connecticut River at Old Saybrook, which comes down about 400 miles from Quebec. Not counting frontage on inlets and estuaries the Sound has about 217 miles of coastline. That's about as much coastline as the state of Maine.

Along the shorelines there are homes, harbors, and beaches, of course, and farther out there are lobster pot floats, bigger boats making passages, tugs with barges, and tree limbs and trunks pulled into the water by waves and high tides. And patches of breeze, tide rips, schools of bluefish and bunker, local fishing boats out on Long Sand Shoal, Five-Mile Reef, the Branford beacon,

Execution Rocks, and dozens of other spots, some plain to see, some secret.

And there's the current to watch—always the current, running east or west, flooding or ebbing, with fast lanes, slow lanes, and back-eddies according to the bottom or the projections of the land. For anyone who runs a boat on the Sound, especially sailors, the state of thc tide and the flow of the current have always been an absolute fixation. With two knots of current going in your direction, life is good and you feel smart. When it's against you, you're in the hands of an evil lunar plot, scrabbling up a slow horizontal waterfall, studying every buoy and pot float to see if the moon is beginning to release its manic grip.

There's a lot of weather to watch, too. Early morning is usually quiet—good for powerboat voyaging. Sometimes the Sound is so flat and calm, and the horizons so indistinct, that it seems as if you're traveling over a gray mirror under a silver sheet. Sometimes you run into pea-soup fog, too, and have to feel your way along, watching the radar and AIS, checking for big commercial

Consider the days when the Montauketts and Niantics inhabited those waters, working the shoreline and its inlets and rivers, taking shellfish, crab, and all sorts of fish and eel by hook, trap, net, spear, and weir. It wasn't so long ago that they were here.

traffic, dreading the yahoo bombing through the cotton at 30 knots without a clue or care in the world.

You have to keep a good weather eye, too. Incoming fronts can kick up a nasty chop, not so bad mid-Sound where it's deeper, but tough along Long Island in a northerly or along the Connecticut shore when there's a hard wind from south to west. In fair weather there are thermal cloud banks over the land, and sometimes line squalls and thunderstorms. And once in a while we get a true black squall that will knock small boats right over and even lay a big displacement boat on its beam-ends and keep her there until it's good and ready to let up.

When you run up and down the Sound you can think a lot about the bottom, all the way from New York to The Race and Plum Gut, where most of the water comes in and out. You consider, first, that the ditch was formed by the runoff of streams and rivers from the north, and that eventually a glacier came down and pushed a bunch of New England southward across the basin and left its moraine—Long Island—before retreating. Today the bottom of the ditch on the north side, at least as far east as the mouth of Connecticut River in Old Saybrook, is mostly mud, created by eons of autumn leaves and the life-cycles of billions of creatures, inhabited by some of the best shellfish in the world and punctuated here and there by rock formations. Some of the rocks rise above the bottom in deep water, providing structures for fish and fishermen to populate. Some are closer to the surface, often in boat-bashing places, so there are plenty of buoys to watch for. And some rocks, along the Connecticut shore, are islands.

The north shore of Long Island is different—most of it is one long, straight run, big bluffs in the middle, with less mud and more sand. There are only a few places to put in—Mattituck, Mount Sinai, and Port Jefferson—until you get as far west as Huntington. But then it's on to the storied harbors of Oyster Bay, Hempstead, and Manhasset, and then Throg's Neck, Hell Gate and the Big Apple itself. Along the way, if you stick close to shore, you can see how a lot of the One-Percenters live. In fact there are amazing houses, harbors, and towns on both shores of the western Sound as it narrows.

And everywhere you go you can think about the people who used to live along the shores—Pequots, Hammonassets, Quinnipiacs, Matinecocks, Manhassets—and their place names. Menunketesuck, Niantic, Housatonic, Seatauket, and on and on. As you watch the fishermen in their center-consoles congregating in The Race and Plum Gut with fishfinders pinging madly, you can consider the days when the Montauketts and Niantics inhabited those waters, working the shoreline and its inlets and rivers, taking shellfish, crab, and all sorts of fish and eel by hook, trap, net, spear, and weir. It wasn't so long ago that they were here.

The Sound today is surrounded by some of the most crowded land on the planet, but it's also remarkably clean, especially compared to the decades of the 1960s and '70s, before the battle against pollution was fully joined. The ospreys, great herons, and egrets are back. We have eagles. We have seals in the winter, and even the occasional whale. But it's also the age of overpopulation, of disposable plastic bottles and mylar balloons, oil spills, fertilizer runoff, drainpipe overflows, and beach litter. One good-sized storm flushes tons of all these things into the Sound. So when we cruise home waters we leave the fish alone and try to take out the trash—same fishnet, different purpose. Chasing down a bunch of balloons scudding across the surface of the water in a fresh breeze is fun, and good for boat-handling skills.

If you have a wicked case of wanderlust but only have the afternoon off, remember the wonders of your home waters. Take a closer look. Those ruby deck shoes are yours. Click the heels together three times…

Watch Out for Whales

Baleen whales have terrible breath. They take big mouthfuls of krill and small fish and whatever else happens to be in their way, then shut their mouths and push the water out through the baleen filter with their tongues, keeping the food inside. When they exhale near you to windward it'll knock your socks off. But if you've been close enough for that to happen, you've gotten a whiff of something truly amazing, and something that most people won't experience.

The whales sounded halfway to us,
increased their speed underwater, surfaced
and blew right behind us, sounded again just aft,
swam under us, turned 90 degrees,
and set off on a new course inshore,
exactly as if they'd been basketball players
using us as pick or a screen.

I've seen whales from the decks of small boats once in a while over the years, and every time it's a gift from the universe: The sperm whale calf resting on the glassy surface 200 miles from Bermuda. Gray whales cruising south inside the Baja peninsula. Humpbacks on Stellwagen Bank. And on two occasions as we were sailing to Maine at night, whales between Cape Ann and the Isles of Shoals surfaced so close that we could inhale their fishy exhalations and listen to them glide alongside us.

So it blew me away when video surfaced of a whale breaching in my home waters of Long Island Sound, off Charles Island in Milford. When I was a kid in the '60s we had pods of bottlenose dolphin around once in a while, before the pollution got too bad. Now that the water is cleaner again the dolphin are coming back, along with seal and porpoise in the winter. And about a decade ago I saw a pilot whale in murky November weather off Faulkner's Island. But now we're talking humpbacks—great whales—full-sized leviathans.

Remember the Dr. Seuss book, *McElligot's Pool*? A kid named Marco is fishing in a little landlocked hole full of junk and old boots. A farmer laughs at him and says he'll never catch anything there. But Marco imagines that the pool is connected to the sea by an underground stream, and that he might just have access to all sorts of fabulous creatures. Well, now when I'm cruising up and down Long Island Sound I'm just as wheezed up as young Marco.

I don't know when the great whales were last here, chasing school fish into the Sound. Maybe before the whalers arrived in Sag Harbor and New London, back in the times when native tribes inhabited the waters around The Race and Plum Gut. Between then and now mankind has almost managed to wipe out several species of whale.

It's a blessing that they're coming back despite a few persistent whale fisheries, and there's increasing evidence that whales and humans can share bonds of consciousness. If you've never seen Amanda Cotton's photos of sperm whales off Dominica, or Tony Wu's photos from the Indian Ocean, or followed Ocean Defender in Hawaii in their interactions with whales off the coast of Maui, it's worth immersing yourself. But it's a curse, too, and a strange, ironic sensation—because with things as they are today, whales shouldn't be in Long Island Sound. Without a doubt, they're ingesting trash at the surface—mylar balloons, styrofoam, baggies, plastic bottles, you name it. And Long Island

Sound, except after a major rainstorm or storm tide, is a relatively clean body of water; whales have it worse elsewhere—even the North Sea, where 13 sperm whales recently washed up on a beach with stomachs full of plastic waste and even car parts.

Along with the surface trash there are lots of people plying the Sound with propellers who aren't used to sharing the water with whales. In fact a humpback was found dead in Lloyd Harbor, Long Island not long ago, apparently having been struck by a boat. The more whales who come here, the more encounters there will be. The osprey and great herons are another matter—they've been back for years—and eagles, too. It's a birdwatcher's paradise near the marshes and along riverbanks. We've accommodated them, and they us. Floridians have accommodated manatees, and it's a thrill to see manatees and heron and dolphin all plying the Intracoastal Waterway with boaters.

But great whales in restricted, boat-infested bodies of water—that's just a recipe for trouble. As much as most of us like them, we have a lousy track record with them species to species, and not just from the times when it was baleen for corsets, spermaceti for watchmaking, and oil for lamps. Modern sonar, engine noise, and dense boat traffic all make for a rotten whale environment. The Federal Marine Mammal Protection Act sets rules for staying away from whales and not bothering them, but most boaters probably aren't aware of the rules, and the whales certainly aren't. They're just following schools of baitfish wherever they might lead.

So it's up to us to avoid them, not the other way around. The trouble is, the great whales are magnificent and mysterious, and we want to see them.

That's what my wife and daughter and I were doing on Stellwagen Bank some years ago in our 26-foot inboard, in the Atlantic just northeast of Cape Cod—looking for whales early on a still, hazy morning. And we found them—two single whales in the space of 15 minutes or so, swimming slowly westward, towards the mainland. As we drifted with our coffee and kept a lookout, two big whale-watching boats, one of which had steamed out from Boston under

the aegis of the New England Aquarium, showed up out of the haze. We were lying about 800 yards away in neutral with our engine idling when the tourist boats throttled up and closed in on a three-whale pod. The whales turned and swam directly toward us with the whale-watchers in pursuit. The whales sounded halfway to us, increased their speed underwater, surfaced and blew right behind us, sounded again just aft, swam under us, turned 90 degrees, and set off on a new course inshore, exactly as if they'd been basketball players using us as pick or a screen.

It was nice to think we'd helped the leviathans foil the aggressive gawker-boats, but of course we were potentially dangerous and bothersome to the whales, too, so it was with mixed feelings that we made our way back inside the shelter of the Cape's hook.

The more people there are in any area, the more animals take it on the chin. Suburban neighborhoods are built on land that has always supported a

> Your mind ticks around the attention clock, checking on sheet and halyard tension, headstay tension, cunningham, outhaul, jib leads, traveler; you're always feeling for rudder angle and trying to reduce drag...

big flourish of wildlife. A bear shows up in the backyard, parents worry about kids, and pretty soon the bear is shot. A gorilla is kept at a zoo, a kid finds his way into the gorilla enclosure, and the gorilla is shot. Now we, with our Yamahas and Yanmars, Verados and V-bottoms, are living in closer and closer proximity to the great whales. They can't be captured and put in aquariums; they can't be tranquilized and moved elsewhere. The only things we can do, and have to do, is clean up our water and give them a decent berth.

A Sailor's Brain Under Power

A lifetime of sailing—offshore, inshore, racing, cruising—gives a person an interesting pattern of thoughts and concerns any time he's on a boat. It would take another lifetime, probably in the care of an analyst or a shaman, to unravel that onboard personality and tease out the individual elements, but quite a few of those elements involve the persistent urge to make a sailboat go faster, or to handle the water better.

Any racing sailor will confirm this: Your mind ticks around the attention clock, checking on sheet and halyard tension, headstay tension, cunningham, outhaul, jib leads, traveler; you're always feeling for rudder angle and trying to reduce drag; you're thinking about heel angle, fore-and-aft trim; you're studying the instruments to see what effect each move has. Then you're thinking about how to get where you're going—when to tack or jibe, the effect of current and leeway, how best to steer through the wavetrain. And all of this is layered on top of the more basic but also absorbing routines of seamanship and safety.

So when we started cruising our 26-foot powerboat many years ago, it took just a few minutes on the first leg of the first run for me to start feeling very strange. We were going 12 knots—about twice as fast as what I was used to in cruising sailboats—and exactly in the direction we wanted to go. (It might even have been directly upwind.) We were level, shaded, and comfortable. We would be at our destination in about three and a half hours, and after those first few minutes I was... well, I was bored. And disappointed, and even sort of embarrassed. There seemed to be nothing to do but steer and watch the gauges and the chartplotter.

My rationale had been ironclad: With work and family life as they were then, it made sense to trade the challenges and pleasures of the sailing voyage for the convenience and speed of the power voyage. I could deal with my sailing jones by racing Lasers and OPBs (other people's boats). I'd understood that power-cruising offered no particular challenge unless the engine crapped out, and required little skill compared to sailing, especially with most of the piloting chores being handled electronically. But I hadn't figured on the boredom. Or the constant engine noise.

All I can say now, quite a few miles later, is that the boredom faded away pretty fast, because the convenience and ease of running a powerboat did nothing to damage my attention clock; it just changed some of the items my mind visits. Other things evolved, too, during the same years. Before the changeover I had always been focused on crossing the water as fast as possible in whatever boat I was in—offshore racer, one-design dinghy, or motorboat. But as the creaks and squeaks of age made themselves persistent and unmistakable, I began feeling more of a closeness to the water itself. I no longer cared so much about speeding across it. I became content—no, deeply happy—just to be on it or in it; to swim and row and idle around, to watch it and learn about it, and help take care of it.

As for the engine noise, there's no way around it. You have to take the bad with the good sometimes. It's not much worse than sails slatting in a calm.

With all that as background, and at the risk of attracting the interest of the mental-health trade, I'm going to give an example of an inner monologue that might occur during a summer run eastward in my local waters—in the mind of a sailor without sailing to do.

Time: 0800, Conditions: Hazy, a few southeasterly zephyrs; calm; slack tide, starting eastward ebb. [Underway from the Thimble Islands, east of New Haven] Good water flow out the exhaust. Check. Wife happy, check. Full mug of coffee, Check. Full fuel, clean heat-exchanger, belts tight, clean bottom, full water. Sweet. Enough food to feed eight people for a month, even though it's just the two of us for four days. Check.

Funny back-eddy on this end of the anchorage. Look at how Dan's boat is riding up on its mooring. Note time. Should make a fast passage. Clear of anchorage, increase to 1300 rpm; let the engine warm up to operating temp. Nudge the throttle a bit to get the transom lip up on the surface. There we go. Jeez, zero boats around. All alone on the Sound. Love it. Still a bit of head current showing on that bell. Turn on VHF to monitor Channel 16.

Man, those are some nasty rocks. Still spooky after after all these years. I need to come up with a better coffee-holder. Engine sounds great. Temp's up to 180. Let's get up to cruising speed... 2000 rpm. Nice... 12 knots over the bottom, so the current must be just about slack. Look at that wake. That's a damn slick wake. I like that lunch hook arrangement, with the coil over the rod holder. Maybe this weekend I'll put an eye-splice through the last chain link. Need to make new fender pendants, too.

Time: 0820. Little bit of a fair current on that lobster pot float. Speed's up to 12.5 knots over the bottom. Used to get 12.5 in slack water, but now it's 12 even. The engine's got another 500 hours on it since those days. The Coppercoat bottom is even slipperier than the old paint. Only difference was the bio-diesel we used to run. That animal-fat bio had a wicked high cetane rating. Wonder if that made the difference? Temp: 180. Good.

Time: 0840. Hammonasset. What a word. Wonder what it means? And how did that huge sand beach form in the middle of the shoreline right there? Where did the sand come from? It must have to do with Meig's Point [looking at plotter]. The point runs northwest to southeast; makes it a lee shore in the prevailing southwesterlies. Would that do it—wind and waves pounding on rock? Or was it mostly just the ebb tide carrying sediment against the point? Need to check that out. Hammonasset. Ham and asset. Ham and eggs. Man, those guys had a long pull in 1777 when they rowed from here to Long Island and then portaged over the North Fork to raid Sag Harbor.

There are the fishermen—half a dozen boats off Six-Mile Reef. A pretty quick run out from Clinton or Westbrook. Blues, striper, blackfish? All of the above? Maybe should keep a rod on board. Nah. Just one more obsession. Temp: 180. Good.

That's a nice center-console there. Man, flag-blue makes just about any hull look good. Maybe a Regulator? Twin Yamaha four-strokes. It'd be fun to have a fast center-console. Nah. Can't cruise in 'em. Can't sleep in 'em... but it'd be pretty cool just to have one to bomb around in... Maybe should paint this boat flag blue when the gelcoat fades.

Time: 0900. [Doing some knee bends] This engine noise is still too much on a long run. No room in the engine box for more insulation... the new mounts didn't help a bit, at least at idle... There go two of those little tugs—Nordic Tugs?—cruising west in tandem. Maybe two retired couples—old friends? Be a nice way to go, nice and slow, sipping fuel, plenty of comfort. Hmmm. We're losing the horizon to the east, and the sun's deep in the haze. Yellow haze. Crap [turning on radar, turning on iPhone AIS app]. Temp: 180. Good.

Time: 0915. Frikkin' fog. Better slow down a bit. Get the horn up here. If I ever have to blow this thing, somebody's gotten way too close. Jeez, why do people still call on 16 for radio checks? Nice and easy. Wife happy? Check. Compare helm compass to GPS course-over-ground. Still pretty accurate, just two degrees off on an east-west course—can't steer much better than that—and spot-on north-south. Not worth trying to tweak it. Wonder if I should run down to the Long Island shore. Nah, might as well stick it out. Better current here, fewer fishermen, and I can see commercial traffic on radar and AIS. Temp: 180. Good. Just have to watch the radar. Maybe should get a better radar someday. None of the connections will fit, though, this thing's so old. Amazing how a $4 app on an iPhone turns out to be one of the most useful piloting tools.

Time: 0945. What makes fog thicken and thin like this? The sun warming things up, sure, but also changes in water temperature underneath? Bits of breeze blowing through? More breeze building now, and look at that big tide line. Foam, twigs, weeds, Mylar balloon, plastic bottle, plastic bag. Let's stop and net some of that crap up [calling wife: "Queequeg, the net, starboard side!"].

Time: 1030. Nice! The fog's lifting just in time for us to squirt through Plum Gut without an Orient Point ferry bearing down on us in 50-yard visibility. Man, it would be so cool to see a 3D view of the bottom around here with the current flowing over it—324 feet, 211 feet, 63 feet, 188 feet, 36 feet. Is there just a jumble of huge hollowed boulders down there? Temp: 180. Good. Let's follow that ferry through. Big rips in here now... short chop. Whoop—15.5 knots over the bottom [steering through the steep gullies and working the throttle]! How the hell can those fisherman deal with rolling from gunnel to gunnel in that rip? Plum Island. Conspiracy theories. Probably be a resort on it someday. Hope they never take down that water tower. Great mark to aim for. Look at that sloop hobby-horsing into the rip. Barely making any way against

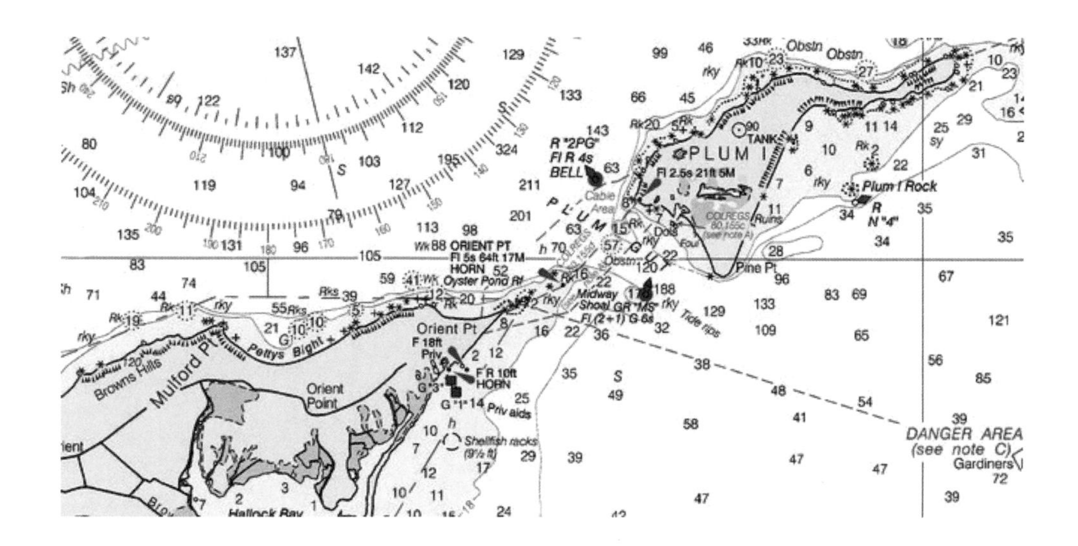

the ebb. Old Ericson... the 3200? Can't mistake that Bruce King cabin trunk. Nice boat. I bet his engine is just about firewalled. Roll your jib out, man! Get a boost! Fog's pretty much disappeared around the corner—I think I can even make out The Ruins over there. Could use an egg sandwich. Or just some gorp. And more coffee. Sweet wake! Maybe should tighten up the steering cable just a bit after we anchor. Temp: 180. Good.

There are no technical means of avoiding
the beat—not in the long run.
There are sensors, but no array of sensors
can match habitual human watchfulness.

CHAPTER 2

Seamanship: Walking the Beat

Ingrained habits of seamanship are simpler, lighter, cheaper, and more reliable than gizmos sold to mitigate the consequences of inattention.

Imagine what it's like for city police as they walk their beat: Check the beauty parlor window to see if Dan is hassling Doreen. See if the Thompson kids are in the alley. Touch the grip of the pistol. Listen for the poker game on the second floor of the hardware store. Note the out-of-state license at the parking meter. Identify the sources of smells; gasoline, burnt rubber, weed smoke, donuts. Adjust the squelch.

What's true for beat cops in the city is also true for sailors. It's the habits of seamanship—the things you check constantly, or have checked constantly for so long that awareness of them is almost subconscious—that are the most important elements of a safe and contented life afloat. These habits are born of a constant, low-grade paranoia that comes from having met Murphy many times. It's not always visible to the casual observer, except maybe as a kind of restlessness. While the casual observer may simply be enjoying the day, with the water hissing by and the sun sparkling, our closet paranoiac is quietly walking the beat—noticing the sheet tension and lead position, thinking about what ports and hatches are open, listening for sounds that are right and sounds that are wrong, watching the weather, coiling, cleaning, restowing, always planning the next move, and planning what to do if that move doesn't work out. If the engine is running, she's checking the gauges all the time, and in case a gauge

> These habits are born of a constant, low-grade paranoia that comes from having met Murphy many times. It's not always visible to the casual observer, except maybe as a kind of restlessness.

is wrong, she's looking over the transom now and then to make sure cooling water is issuing forth. If the going is boisterous, he's lifting the bilge covers for a look once in a while, and making sure things are stowed right, so they won't come flying across the cabin. She's checking the gas valves, the head seacock. He's making sure the heavy-air jib is on top of the pile, that the boathook is lashed securely, that the Lifesling can be deployed without a snarl, that no one's sitting to leeward of the traveler car…

There are no technical means of avoiding the beat—not in the long run. There are sensors, but no array of sensors can match habitual human watchfulness. Take the matter of battery charging, now one of the most arcane issues in boating, though it need not be.

The problem is in understanding, then choosing among, the vast array of "smart" chargers and regulators, dual-output alternators, high-output alternators, doubled alternators, battery combiners and isolators, eliminators, smart inverters, dumb inverters, extroverted inverters, and any conceivable combination of all the above. Cripes.

Some of these pieces of gear are complex, some aren't. The real trouble is that they not only vary in use and efficiency, but overlap each other in their capabilities. The boatowner is presented too complex a mix of options, not even including how the stuff is made, mounted, and maintained. And few of those items are fixable on board. There are specialists who will think through these matters with owners and install the gear. But the gear will malfunction, of course, eventually.

I prefer the hassle of paying attention to the charge indicator every time I go by the panel. It's reflex now, like glancing in the rearview mirror. And that simple isolator switch—I'd miss exercising it manually, even considering all the terror it caused me when I was a kid and I was threatened with summary execution if I turned it the wrong way and fried the diodes. And all those requests: "Honey, can you turn the switch to BOTH?" What would we talk about? What would become of all the ancient switch rituals? As a sage whose name escapes me said, "Man makes the habits, and habits make the man." It goes double for cops and sailors.

Stowing and Squaring Away

They say there's more than one way to skin a cat. Those of us who would forgo the chance to verify this assertion can still see how people might be tempted to think it would apply in the matter of stowing a boat. But the devil-may-care views that may prevail in the world of cat-skinning don't hold water in the world afloat. The right way to stow a boat—sail or power—is to keep it light, simple, uncluttered, ready for heavy weather, and as free as possible of all the stuff that gets in the way more than it helps. The boat will perform better without extra weight; the crew will have elbow-room, and cruising will be safer.

Some allowances must be made for hobbies and avocations. For example,

CHECKLIST FOR AN UNFAMILIAR BOAT

EXPERIENCED DELIVERY SKIPPERS and charter-boat sailors develop habits over the years that serve them well when they get aboard an unfamiliar boat for the first time and want to make sure they won't be miles away from service or safety if they discover a problem. One key to good habits is the simple checklist, which can be expanded, shrunk, or fine-tuned at will. Here's an example:

WATER PATHS

__ Locate all through-hulls.
__ Check seacocks for operation, double clamps, supple hose.
__ Check stuffing box for leaks.
__ Check bilge for water, oil, general condition.
__ Manually check bilge pump floats. Pumps turn on OK?

STEERING SYSTEM

__ Check for smooth steering, any slack or friction in cables, quadrant, hydraulics.
__ Rudder OK from stop to stop?
__ Rudder bearings and stuffing box OK?
__ Wheel lock OK?
__ Location of emergency tiller and set-up procedure.

GROUND TACKLE

__ Main and auxiliary anchors all rigged?
__ Rode untangled and ready to pay out?
__ If there's a windlass, does it work?
__ Will you be able to let out and haul in the rode manually?

RIGGING AND SAILS

__ Stays and shrouds set up more or less in tune?
__ Plenty of threads visible in turnbuckle screws?

__ Cotter pins present and not dangerous to crew?
__ No meat hooks in wire rigging?
__ Location of all sheets, halyards, and leads.
__ Location and operation of reefing lines.
__ Know how to raise, lower, furl, and reef sails.

SAFETY GEAR

__ Location and condition of flare kit, PFDs, and first-aid kit.
__ Location and condition of flashlights and spare batteries.
__ Location and contents of tool kit—got screwdrivers, wrenches, pliers, a hammer, electrical tape, duct tape, rigging knife, boathook?
__ VHF radio check

DINGHY

__ If it's an inflatable, is it fully inflated? If a solid dink, no leaks?
__ Does the outboard start easily and pump cooling water?
__ Fuel tank full?
__ Pump or bailer present?
__ Painter not too frayed or too short? Bow eye secure?

AUXILIARY ENGINE

__ Check oil, coolant, belts and filters at the dock.
__ Fuel tank full?
__ Run engine, familiarize yourself with controls, gauges, and fuel shutoff.
__ Check for cooling water from the exhaust (and check it often).

BATTERIES AND ELECTRICITY

__ Check battery terminals, electrolyte levels, state of charge.
__ Location and operation of main battery switch and circuit breakers.
__ Go over switch and breakers with the crew.
__ Check nav and cabin lights.

HEAD

__ Check toilet operation with crew.

__ Pump works easily and effectively? No leaks? No foul odor?

__ Holding tank fill level OK? Valves operate OK?

__ Got toilet paper?

FRESH WATER

__ Check pressure-water system for operation and condition of water.

__ Water tanks full?

__ If they need to be switched, know how?

__ Is there a manual water pump in the galley? Nice to have.

REFRIGERATION

__ Check icebox and components for coldness.

__ Get information on engine charging schedules and post in galley or at breaker panel.

GALLEY

__ Ample food

__ Ample potable water

__ Eating and cooking utensils

__ Pots and pans

__ Dish soap

__ Food storage containers

PERSONAL ITEMS

__ Adequate warm clothing and foul-weather gear

__ Medicines, glasses, contact lens gear

__ Snorkeling gear

__ Towels

__ Sunscreen

__ Ziploc bags

The devil-may-care views that may prevail in the world of cat-skinning don't hold water in the world afloat. The right way to stow a boat—sail or power—is to keep it light, simple, uncluttered, ready for heavy weather, and as free as possible of all the stuff that gets in the way more than it helps.

there are a lot of people whose boats serve mainly as floating electronics platforms. Give them a nice spring day and they're just as happy configuring WiFi adapters or tuning instrument systems as being out on the water. These people will need a bit of extra space for terminals, wire, and meters. Others might absolutely need the space for Buddhist prayer candles. As long as gear is used regularly and it contributes to enjoyment, it belongs. But there's no sense in carrying too much "just in case" stuff. If you try to take along gear to cover most of the breakdowns or entertainments you can imagine on board, you'll overwhelm the boat. Take prudent gear and fun gear, but trust your habits of seamanship to see you through, and the boat, your shipmates, and the surroundings to entertain you.

The same goes for carrying gear that you intend to install someday, but are avoiding because it requires drilling holes in the deck or rearranging a breaker panel or removing a bunk bottom. That gear is better stowed in the garage or a closet ashore, where it's safe and dry, until the time comes when you can really concentrate on it.

When it comes to outfitting a new boat or re-stowing an old one, it helps to make a simple chart consisting of a blank sheet of paper with a heavy line down the middle. On one side of the line are items of necessity. This is gear that is used in normal operations, plus the equipment and spares that allow you to leave Point A, run into a nasty streak of bad luck, and still arrive at Point B safely and without assistance. On the other side are items of convenience, including a token number of things that take into account your predilections—yoga mats, pink flamingos, whatever makes you happy.

Many people who have been around boats for a long time become neater and more organized, some to the point of compulsion and obnoxiousness. But there are good reasons for it. First, boats don't have much room. If you're traveling with four people in a space the size of camper, things can get out of place almost instantly if everyone isn't in the habit of keeping their belongings squared away. This causes friction and wastes time. Second, there's a safety element in play. If you can't find a pump handle or a fuse or a flashlight because it's been buried under other ill-stowed gear, then whatever problem you need that item to address is going to be compounded.

Here are some general stowage tips for the cabin:

- Every crewmember should have a drawer or locker or hammock of their own. If there isn't enough stowage space built in, they should operate out of their duffel bags. Duffels should be kept zipped up and out of the way as much as possible.
- If there's no wet locker, crewmembers should be told where to hang their foul-weather gear.
- Put things that are needed often in the most accessible spots, heavy things low and inboard, light things higher and outboard. Figure out what might hit someone in the head if the boat goes on her beam-ends, and move or secure it.
- Consider carefully where to stow and mount safety gear, and keep it instantly accessible. People regularly mount fire extinguishers where reaching them would mean extending an arm through a galley fire or engine fire.

> If you can't find a pump handle or a fuse or a flashlight because it's been buried under other ill-stowed gear, then whatever problem you need that item to address is going to be compounded.

So, boatsense includes an understanding of what to have on board, no more and no less, and where to put everything so that you can find it when you need it. It's true that Murphy's Law comes into play: The one thing that you decided you wouldn't need is exactly what you do need—but that's OK. If it were that important, you would have brought it; since it wasn't so important, you didn't really need it. If you still think you need it, you can go out of your way to get it, or bring it along next time. But next time you won't need it.

Stowage logic, orderliness, and cleanliness all go together. It's a matter of crew comfort and safety—same as on land, but more so. Let us speak, for example, of oily crud in the bilge, which is an abomination and a danger. Years ago we went on a short cruise when my son and his friends were about seven, an age that involves lots of running up and down the companionway and lots of mess. After the first couple of days, when Uncle Dave and I got tired of sweeping up the broken crackers and little plastic juice straws and wrappers and other detritus, we pulled up the bilge cover and said, "Have a look down there. See that pump? Well if you're not neat, and you don't clean up your mess, those straw wrappers are going to clog that pump, and then if we spring a leak and start taking on water, the pump won't work, and the water will rise and rise, and finally the boat will sink and then... we'll die!" We followed that with diabolical laughter, and our lesson actually worked for a few minutes.

It wasn't just a ruse, though. If cracker crumbs and sawdust from work projects and small oil drips aren't cleaned up religiously, they do end up in the bilge. Bilge sludge stinks, clogs pumps, and can be a fire hazard. And because it's nasty to deal with, it's often ignored, which of course makes it worse.

In warm temperate climes or the tropics, traces of food left out in the galley or the cockpit invite critters ranging from spore-size to bird-size, with bugs and rodents in between.

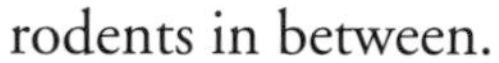

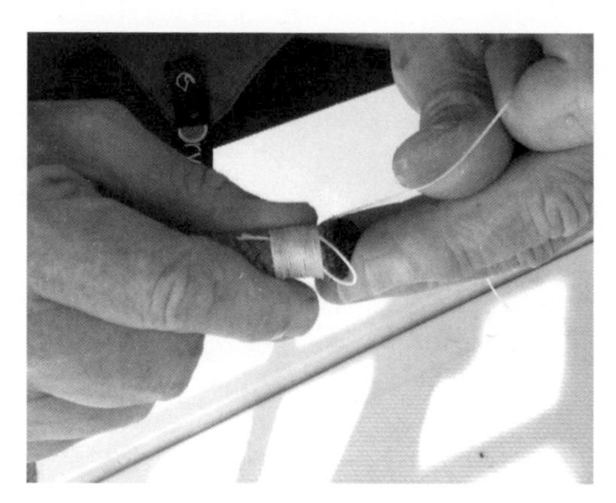

In the old days, crews of navy ships swabbed and holystoned the decks every morning, not necessarily because the decks were dirty— they rarely were— but because it was part of a ship-keeping ritual that helped imbue the sailors with a sense of responsibility for their vessels. Painting, tarring, splicing, sewing, mending netting, and dozens of other maintenance jobs were done ritually and early, before there was a broken line, a blown-out sail, or rust. These rituals meant that on or off watch, sailors had a mentality of watchfulness, maintenance, readiness.

Personal Safety: Training, Process, and Practice

Good seamanship is a state of mind that aims to keep things from going wrong, but things nevertheless do go wrong sometimes, despite good preparation, and matters get very serious when someone's health or life is threatened. Injury, drowning, hypothermia, sudden illness—any of these things can happen on boats, and the farther the boat is from land and organized help, the more dire the problems. So if taking a calculated risk is one definition of adventure, then any powerboater's or sailor's calculation, every time they go out on a boat, should include systems and plans for personal safety. This subset of seamanship requires some formal training, and luckily there are courses available through the U.S. Power Squadrons, US Sailing, BoatUS, the American Red Cross, the Cruising Club of America, and other organizations. In book form, the best reference is *The Annapolis Book of Seamanship*, which John Rousmaniere keeps updated with the latest in safety-related wisdom.

By "requires" I don't necessarily mean a legal requirement; I mean that any new boater really should take a formal course just to get a basic idea of what's involved. Even then, no reading or classroom chalk-talk can take the place of thinking through safety processes yourself, on your own boat, and practicing (really, no kidding, hands-on practicing) the actual moves necessary to preserve life and limb.

The thought of someone falling overboard, for instance, should of course bring to mind proper-fitting personal flotation devices (PFDs) for everyone aboard. But that's only a fraction of what needs to be considered. What if the person in the water is injured or unconscious, or both? What if it's dark out? Who's going to keep an eye on the person? (Someone who falls overboard in anything but calm water can be very hard to spot even a few dozen yards away.) What's your procedure for stopping and turning around and returning to the person? It may not be too difficult in an outboard runabout, but what if you're offshore in a 40-foot sloop with the spinnaker up? What if it's just you and your spouse on a 30-foot cabin cruiser? And when you finally get the person alongside the boat, how are you going to get them back on board? Get your training, have a plan, and practice it. Chances are, you'll need a better plan. And more practice.

Any boatowner should have up-to-date, comprehensive first-aid training, and of course a decent medical kit aboard if the boat travels more than a few minutes from land-based help. This is especially important if there are going to be others on board: The captain is responsible for everyone on the boat. Emergency medical training, like being able to tie a bowline or knowing how to find north, is handy on shore, too.

One more thing: when you take anyone out on your boat, make sure to outline your safety procedures for them—how to use their PFD, how to stay on the boat, what's expected of them in case something goes wrong, how to stop the boat safely and what to do if it's *you* who gets hurt or falls overboard, and how to use the VHF radio. It's also good to keep people aware of their location, so that if they have to make a call on the VHF or on their cell phone, they can tell rescuers where you are.

CHAPTER 3

Embrace the Hacksaw

Boat maintenance and repair are inevitable—and not in small amounts. If you can learn to enjoy these chores rather than dread them, you'll save money, and you'll learn skills that are even more valuable.

When I see the ads for gorgeous boats zipping around on placid waters or sitting in idyllic coves with really good-looking people on them, I have two equally powerful reactions. One is: "What a lot of hooey! That supermodel won't be so super when she's crashing around in three-foot chop getting her fillings knocked out. And look at that puny little [plastic, cheap brass, pot metal] doo-hickey. I'll give it about a month of [salt water, direct sunlight, physical abuse] before it turns [yellow, green, brown] and craps out."

The other reaction is: "Man, I want that boat." Depending on the boat in the ad, I want it for diving, singlehanded sailing, marine rescue, harbor cruising, long-distance cruising, power-weekending, or fishing. And I don't even fish.

Both of these reactions are valid for the experienced boat addict, because boating always brings with it two powerfully opposed yet complementary elements: Agony and Ecstasy. Yin and Yang. Triumph and Tragedy. Thesis and Antithesis. You get the picture.

Maybe that's what makes boats so compelling. They're not trivial. Not a single one of them. Not a superyacht, not a kayak. They are adventure, and commitment, and joy. They keep you up at night dreaming and fantasizing and planning. They get your senses all flared up.

But let's take a step back and consider what that "negative" side of boating—the "yin" side— really means. It's too often swept under the rug by people trying to sell boats and promote boating when, in my opinion, it should somehow be used as a selling point. Because it's there, it's part of the game, and it's not going away.

If you want real, lasting enjoyment as a boat owner, you'll need to get there by one of three paths:

1. Bring a good set of skills ranging from fiberglass maintenance to engine mechanics.
2. Have a lot of friends and mentors with those skills, and be a good student.
3. Arrive rich.

But even if you arrive rich and can pay others to do the work, you'll still be better off getting your own hands dirty. Because if you can't learn to love hacksawing in a cramped space with skin coming off your knuckles at every stroke and sweat pouring from your brow, you cannot really love boating as much as it can be loved. This is the sad truth about those who pay others to do the work. They're missing much of the fun—because running a boat that you maintain yourself is a whole lot more enjoyable than running one that has been maintained by someone else's concentration, effort, dedication, and understanding.

When you do your own repair and maintenance work, you'll find that the ratio of work to carefree boating will be, at best, roughly 1:1. Put another way, for every hour you spend with things running pretty well and your spouse all smiles and the kids having a blast, you're going to be working on the boat for at least an equal amount of time. Sometimes the ratio is more like 3:1 or worse, especially when you're working on a fixer-upper.

> Running a boat that you maintain yourself is a whole lot more enjoyable than running one that has been maintained by someone else's concentration, effort, dedication, and understanding.

As for knowledge and skill, it doesn't matter where you start on the spectrum—there's always someone smarter and more skilled than you, and all the way up the chain the smarter ones (at least all the ones I've known) are willing to share their secrets with you.

But you have to be willing to learn. Maybe there are two subsets of people who can't screw in a light bulb—one who says, "Oh, I can't even screw in a lightbulb. I don't really *do* lightbulbs." And one who says, "I can't even screw in a lightbulb. I think I'm going to keep trying."

When you finally screw in that lightbulb successfully you get a feeling of satisfaction. The next time it's easier. With momentum comes confidence. And, where boats are concerned, that's when you start breaking things that are a lot more expensive than light bulbs. So you regain your humility and just go on learning and busting stuff along the way, but gradually getting a little more competent all the time.

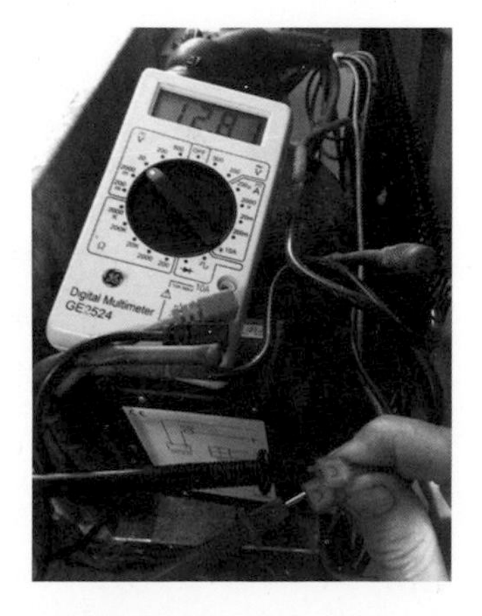

When experienced boat people hang around together, they discuss hacksaws and hose a lot more than they talk about the supermodels they've had lounging around the cockpit—although I'm sure that would be everyone's preference. They talk about alternators and epoxy and stuffing boxes and exhaust elbows and sea strainers and joker valves. If you don't know what a joker valve is, picture the lips on someone giving you the raspberry. But what's coming through the lips of the joker valve isn't air. It's a lot worse than air.

And another thing: A lot of the fun in boating has to do with the stories you collect, and you can't have a good collection of stories without some maintenance adventures. But maintenance yarns are really boring if you don't do your own work.

"We were in the middle of the channel and the engine died. We got a tow in. Then I called my mechanic."

If you don't know what a joker valve is, picture the lips on someone giving you the raspberry. But what's coming through the lips of the joker valve isn't air. It's a lot worse than air.

Is that a good story? No, that's a boring story. But I could tell you a story about changing a joker valve while heeled over offshore in rough weather that would make you laugh and cry and throw up all at the same time.

Entropy rules the universe, and it is the nature of all things to break. Thanks to the environment they operate in, it is the nature of boats to break more often and in less convenient places. Because preventive maintenance and unexpected breakage are both inevitable, you might as well come to terms with them and do your best to transform them from pain to pleasure. Then both the maintenance that prevents breakage, plus the repair of the breakage that comes anyway, are challenges that are satisfying to overcome.

There are also some problems on boats that come from poor design, poor installation, poor materials, or some combination of all of the above. Those things have to be fixed permanently, replaced with something better, or taken off the boat for good—because they will never stop haunting and annoying you.

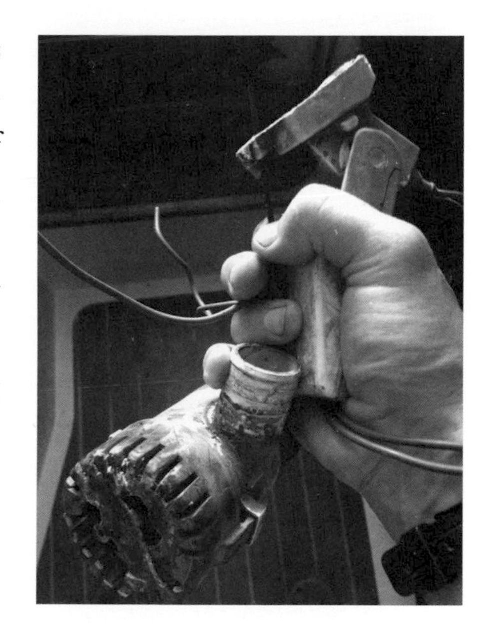

Other problems come from things being installed on board that never should have been there in the first place. I won't be specific here because maybe you like plumbed-in air-conditioning on your boat and I would rather have a couple of good hatches and 12-volt fans, and we're never going to agree. If you're OK maintaining the extra through-hull valve, the strainer, the pump, the wiring, and the generator or extra battery power to run it when your shore-power umbilicus is removed, good for you. I'm OK with not being perfectly comfortable.

Another example. I bought a very used boat with a 15-gallon holding tank and an electric toilet with a macerator. The contents of the holding tank had long ago turned to dry pellets. The hoses were all clogged and needed replacing. The macerator was clogged and dead. The only thing that worked was the electric head pump. It made a sort of screeching noise as if it was frustrated with life, and had been for a long time. With great sweaty effort (and, incidentally, my hacksaw) I pulled the whole miserable system out of the boat and replaced it with a five-gallon self-contained toilet, because that's all my family needs, and I'm perfectly willing to empty it ashore. Since nobody else has to do it, everyone's happy.

Whenever possible I get rid of bothersome things and move toward simplicity. I discovered long ago that removing the offending item entirely and then replacing it with something less fancy and easier to fix usually works for me.

VOLTS, WATTS, AMPS, AND OHMS

IF YOU'RE NEW TO BOATS, or even if you've been around them for a while but have been avoiding the electrical bits, sooner or later you're going to need to become more familiar with DC electricity than you are if all you know how to do is jump-start your car. To help get you going, here are some very basic definitions you'll need to understand.

DC means Direct Current. It's the kind of electrical current produced by batteries. Batteries found in cars, trucks, RVs, and boats are almost always 12-volt DC. Flashlight batteries are DC, too. In fact you can create 12-volt potential by linking together just eight 1.5-volt flashlight batteries—AA, C, or D (1.5 x 8 = 12).

Fuses are rated for a certain maximum amperage flow, above which they will 'blow,' interrupting the circuit. Amp ratings are shown on the fuses above. Circuit breakers work the same way but 'trip' when they're overloaded.

Here are some basic 12-volt definitions. **Voltage/Volts** = the amount of potential energy available to push electrical current. Since electricity is invisible, it's convenient to picture voltage as the potential pressure in a water system. For a battery-powered system, think of a water tower with a big tank on top. It's drained by gravity, and the way water flows out of it depends partly on the volume and weight of the water in the tank at any given time (the "voltage" equivalent), and partly on the characteristics of the drain pipe. In an electrical system, wires and other conductors are the equivalent of water pipes.

Amperage/Amperes/Amps = the flow of electrical current through conductors like wires. Think of it as the amount of water flowing past a single point in a pipe at a given time.

Wattage/Watts = the amount of energy expended, or used. Think of it as the water needed to fill a glass (a few watts) or a swimming pool (lots of watts).

That's as far as the plumbing analogy goes, though, because unlike water systems that run from source to drain, electrical systems run in "circuits," in other words in a circle, from power source to usage ("load") and back to the power source, with switches and fuses or circuit breakers in between to interrupt the flow of electricity as necessary.

Ampere-hours (Ah) = the current in amperes multiplied by the amount of time it flows. Batteries have ampere-hour capacity ratings that give a general idea of how many amps can be drawn from the battery for how long. In a perfect world, a battery rated for 90 amp-hours would be able to give you 90 amps for one hour, 45 amps for two hours, one amp for 90 hours, and so on. In reality, you can and would use only a portion of those amp-hours before the battery should be charged again.

Ohm = a measure of resistance in a wire or other conductor. Resistance is determined both by the wire's length and its thickness, or gauge. The thicker the wire, the more easily current will flow through it. Resistance always creates heat, and the greater the resistance, the more heat. Try to put too much current through too small a wire, and you can create enough resistance to start melting things and causing fires. This can happen even in a simple 12-volt system, so always use common sense and generous wire gauges.

In most cases it's unrealistic for a boatowner to measure for ohms, simply because boat gear manufacturers don't usually offer a baseline of resistance to measure against. Instead we measure for an abnormal drop in voltage in a circuit, which would indicate corrosion, too small a wire size, or a poor connection.

Now, here are some of the easiest equations you'll ever have to use:

Volts x Amps = Watts (example: 12 volts x 5 amps = 60 watts)

Watts / Volts = Amps (example: 60 watts / 12 volts = 5 amps)

Amps x Time = Ah (example: 3 amps x 5 hours = 15 Ah)

If you read the owner's manual for any piece of electrical gear, or the stamped information on the gear itself, you can usually discover what it needs for energy input and how much energy it uses. Most marine gear, whether a chartplotter, a bilge pump, a windlass, or an electric windshield wiper, will tell you its current draw in amperes. Then it's a matter of arithmetic to find out if your 12-volt system can handle the task, and for how long. (Note: there's also marine gear made for 24-volt systems, but those are usually on larger or specialized boats; the great majority of small pleasure-boats use 12-volt systems, and in any case the DC principles are the same.)

Working with 12-volt systems is easy, and relatively safe, compared to 120-volt AC. But 12-volt systems are far from benign, for several reasons.

First, a 12-volt battery can deliver a whopping big discharge of current all at once. That's how a 12-volt battery can run a starter motor to get a car or a boat running. That discharge of current can be violent, as you've noticed if you've managed to bridge the terminals with a wrench or a wire. If you have a metal watch strap or a ring on, and that metal becomes part of the short circuit, you can get hurt. If you've used jumper cables, you've probably seen a good-sized spark as a clamp comes in contact with a terminal. So be alert to the danger of short circuits and sparking.

Second, when you produce a spark in the wrong environment, you can cause an explosion. Batteries produce hydrogen and oxygen when they're charging—no problem if the battery is well-ventilated, but a potentially explosive mixture if the battery is in an enclosed, poorly ventilated space or container.

Third, inadequate wiring, unfused wires, poor connections, and corrosion can cause electrical fires, which in turn can catch other flammable material like cloth or paper on fire. If you get involved in a 12-volt installation on your boat, follow the instructions, use big enough wire with short enough wiring runs; use fuses where necessary, and never run wiring under or through any flammable material.

Fourth, the liquid electrolyte in flooded-cell batteries is mostly sulfuric acid, which will eat your clothes and burn your skin, and can blind you if you get it in your eyes. When inspecting and testing the flooded cells in your battery it's a good idea to wear glasses or other eye protection, and rubber gloves. When you pry off the cell covers with a flat-bladed screwdriver, do it gently and carefully, and make sure the battery is on a stable surface. If you get electrolyte on you, rinse with fresh water.

> Entropy rules the universe, and it is the nature of all things to break. Thanks to the environment they operate in, it is the nature of boats to break more often and in less convenient places.

At first it might have been that I wasn't good at repairs, but I also realized that there were certain things on boats that I didn't like because they took my time away from what I did like.

It has also gradually dawned on me that there's no harm in waiting a bit before surgery. Having made many mistakes either by being in too much of a hurry, or by tackling a job with a sort of grim, over-earnest desire to make a "professional" job of it, I've drilled and sawn a lot of things over the years that should have been left alone. Now I've adopted "proof of concept" as a first move, and take a look at things in place before getting out the heavy equipment. It's amazing how many times a thick rubber band, or Velcro, or some waxed thread or marline will take care of something that I'd thought might need a stainless steel or fiberglass solution.

Life aboard boats is definitely beautiful, but rarely in the ways shown in the ads. And it's definitely fun. But it's not always easy, especially when you're new to it. And it's not usually very comfortable. And a lot of the beauty comes from it not being so easy. It takes hard work and setbacks, but if you stick with it those things yield rewards that are much better than what glossy ads can ever show—rewards like skills, understanding, and self-reliance, which are handy both on and off the boat.

Silence of the Engines

The Westerbeke 4-107 had been through three owners by the time I got it. It had plenty of hours on it, and somewhere along the way it had spent time with a lot of water on it, and maybe in it. It had been given bad fuel more than once. It leaked here and there. It was a very needy engine, and I wasn't able to provide for it, either mechanically or financially.

I ran it for three years, detesting it more every day, and fantasizing about a brand-new engine that I could coddle with perfect fuel and lots of air and new filters and regular oil changes and all that. But it went the other way—I finally gave the remains of the Westerbeke to a diesel mechanic in exchange for his last bill, and for his agreement to wrestle it out of my boat, which was also my home at that point.

I went without it for about four more years, and life was simpler and quieter than before. To get in and out of harbor when there was no wind I used a long sculling oar made of stairway railing sleeved in PVC pipe, with a big plastic blade bolted in a notch at the end. It worked pretty well—I could get the boat up to about a knot, maybe a bit more, if there was no head current. I learned a lot in those years about light-air sailing, tidal current, and anchoring. Lots and lots about anchoring.

Eventually I sold the boat to a guy who didn't mind that it had no engine. He sailed it from Connecticut to Florida like that. In the years since, I've spent a lot of time on boats with and without power, intentionally or not, and have more often than not found that the time without power was better in most ways than the time with.

Places with warm weather and predictable breeze make it easier to go engine-free. Years ago my family of four spent two weeks aboard a 30-foot sloop in the British Virgin Islands. We were vacationing,

To get in and out of harbor when there was no wind I used a long sculling oar made of stairway railing sleeved in PVC pipe, with a big plastic blade bolted in a notch at the end. I learned a lot in those years about light-air sailing, tidal current, and anchoring. Lots and lots about anchoring.

but also helping the owners, close friends, get the boat in shape for chartering.

The boat had belonged to a resort, where it had been sailed very hard for 11 years and then mussed up in a hurricane. By the time we got there, our friends had put the boat in good condition. The easy work remained—we set up the running rigging, screwed in a few loose bits, tried to do no harm, and ended up in a pleasant state of lassitude. The only part that got left out of the fix-it plan was the engine, which lacked a wiring harness and control cables.

The lack of auxiliary power wasn't a transportation problem, since we always had plenty of wind to move us around. We also had an inflatable dinghy, so were able to get ice and provisions easily. We ate ashore with our friends about half the time. We each had two or three full showers at the dock during those two weeks, but usually just rinsed the salt off on board at the end of the day.

We were on a mooring in a designated anchorage, so we didn't have to set an anchor light. We had a candle in the main cabin, and we ran electric lights sparingly for dishwashing and reading at night. It was surprising how little electricity and water we needed: In two weeks we ran down one battery, traded it for a fresh one, and probably used about 120 amp-hours of power in all. We refilled the water tank once, and used maybe 80 gallons in all. I'm convinced we could have done the whole thing on less than one battery, but there were no manual or pedal pumps for fresh water. We took to filling three one-gallon jugs at once so that we could keep the battery completely off most of the time and not keep cycling the pump. (I'm not sure it saves electricity to run a pump steadily for 80 seconds instead of in 20 four-second bursts, but that was my assumption, and in any case it made the boat quieter.)

We were a happy ship, neither mutinous nor mangy. The experience reminded me of how little electrical power is really needed for the basics afloat. Where we were, a solar panel could have kept the battery topped up, especially if we could have pumped water by human power.

On extended cruises, refrigeration may become more important for some, but for day-sails, weekends, or short passages ice, as Robert Frost noted, will suffice. True, it can be hard to find in marinas these days, but you can make your own block ice at home.

I wouldn't necessarily advocate an engineless way of life if you have a choice in the matter, and live in a place where the wind is fickle. The main problem is that you often can't get places on time. Sometimes it's not a matter of hours, but of days. This just doesn't work well for spouses, friends, and employers. There are also times when you wish you could get out of the way of ship traffic a bit faster. And there's often no way to predict just when you're going to have an adventure, or find out something new about the boat or the wind or the current. These adventures and learning experiences come at inconvenient times. But they do arrive, without fail, much more often than they do if you have a working engine.

There are some good procedures and habits involved in conservation aboard an engine-free boat. Practiced long enough, they can evolve into a state of mind that serves well both on board and on shore.

MARINE ENGINE CARE

IT TAKES A GOOD MECHANIC to make an abused or neglected engine run well again, but it doesn't take much mechanical aptitude to keep a healthy engine running smoothly, whether it's an outboard, inboard, outdrive, gas, or diesel. The key, obviously, is maintenance, which is a lot simpler than repair.

There are three elements to the approach. First, get over your fears. You don't want to be stuck at the dock, or worse, away from a dock, because you have a phobia about wrenches. You're not going to be involved in deep surgery, so don't tell yourself that you always need a specialist. And don't let fear masquerade as snobbery or disdain for the life of oily fingers and barked knuckles. That's part of the game. Second, pay a few minutes of attention to your engine every time you use it, and know how to troubleshoot the basic systems. Third, be organized about it. Keep notes on what filters you need, what wrenches, what shaft zincs, what fuses. If you work a little bit to keep a decent engine in shape, it will reward you with good performance for thousands of hours. If you neglect it, though, it will eventually let you down, almost certainly at a very inconvenient moment.

Here are a few tips that should keep you away from the mechanic for a long time:

- Keep the engine, the engine box, and the bilge below it clean and as dry as possible. Loose dirt, oil, and fuel are dangerous things to have in the engine compartment. Any dirt that enters the engine, especially the fuel system, will give you major headaches down the line. And with a clean engine space you'll be able to spot problems like fuel and oil leaks and hairline cracks more easily. A grimy engine invites neglect.
- After you've read the engine manual, mark the pages on regular maintenance with tape for a thumb index. Eventually you'll be able to make your own list, and maybe after that you won't need a list at all. But be organized.
- Burn clean fuel. This can't be overemphasized. This means putting in

fuel from trustworthy sources, filtering it carefully, changing your filters religiously, and being on the lookout for any contaminants around the engine compartment or fuel-fill site. If you're in doubt, let paranoia rule. Use a clean funnel with fine mesh or a Teflon screen. You can even pump a pint into a clear glass jar and inspect it before you put it in the funnel. Keep absorbent pads nearby, use them to catch anything you don't want to put in the tank, and dispose of them properly.

- In gasoline-powered engines, ethanol-laced fuel can cause major problems. Make sure to use the proper additives, filters, and water-separators, and monitor your tanks and fuel lines for signs of corrosion.
- If you're running a two-stroke outboard, stick close to the manufacturer's specs for fuel-to-oil mixtures. If it's a four-stroke outboard or inboard gas or diesel engine, change your engine oil religiously. The key is to develop your own system for doing it quickly and cleanly, so that it's not a chore you avoid.
- All engines need plenty of air, both for fuel combustion and to dissipate heat. Make sure your air-filter element is clean, and that your engine space has a free flow of air. A lot of engines now are well-insulated to trap sound and heat, but this is bad for air intake and heat exchange. Keep the engine well-ventilated, especially if you're running for long periods in hot weather.
- If you're running an inboard engine with cooling water pumped up through strainers from underwater, be vigilant about those strainers, both the outboard and inboard ones. It's very common for weed or plastic to be sucked against the boat, starving the cooling water system. This can wreck a water pump impeller in short order, and ruin your day. Check your exhaust water flow and water temperature gauge often.
- Before startup, check your oil level, coolant level, and belt tension for water pump and alternator. It only takes a minute. Regularly check hoses, hose clamps, transmission oil level, and inboard strainer contents.

CHAPTER 4

Buying In

It helps if you're not some sort of swivel-necked Lothario, panting after every transom that glides by. It helps if you have a calm, sober, longstanding view of what kind of boat suits you, and what you plan to do with it. I have a friend who fell in love early with the Westsail 32. This was a few years after Robin Knox-Johnston bobbed around the world in *Suhaili*. It didn't matter that *Suhaili*'s speed made good was about four knots, or that RKJ would have preferred going in a much faster boat—this other fella loved the salty look and sense of security exuded by the heavy teak double-ender, and when the Westsail was introduced, that was that. Now he has one, and he's happy. Never looked at another boat.

It helps if you disdain fiberglass and enjoy nothing more than strapping on kneepads and spending a long weekend with a short piece of sandpaper. It helps, for that matter, if you disdain wood, or lead keels, or L-shaped settees. Any strong belief or conviction that will eliminate candidates or whole classes of candidates, can make these matters simpler. I do not, alas, fit this stern profile. I keep a folder unabashedly called "Boats I Want." It includes a list of designs varied enough to suggest a certain fickleness of interest, if not a mild clinical problem: There are sloops, cutters, and yawls;

multi-outboard center-consoles, single-diesel lobster boats, flashy pod-drive speedsters, Herreshoff beauties, old CCA-era racer-cruisers (almost anything Rosenfeld photographed in about 1958), Dick Newick's early plywood trimarans, and, for chasteness of spirit, several gorgeous pulling boats.

How would it be, I wonder, if I were sailing along at six knots on my Ohlson 38 in a crisp fall breeze, with a kettle simmering on the stove and the morning sun glowing through the mainsail, if a Dragonfly trimaran came tearing past at 16 knots in a welter of spray, with the crew whooping and hollering? Would I be envious? Sure, I would. But then I put myself in their shoes—everything on board is soaked, I have to trim the chute constantly, because 14 knots feels dead in the water compared to 16, and as we pass that gorgeous Ohlson I catch a whiff of fresh-brewed coffee. Some of us will never be happy. It's a curse.

Obviously there's a big difference between the impulses involved in wanting boats and the realities of having them. A good match is certainly possible, but only when we can answer the three following questions with a clear mind: 1. Where are you going to go? 2. Who are you going to go with? 3. How much time do you have? Each of those questions opens to reveal qualifications, provisos, and rationalizations. These are booby traps. The way to spot them is to try the word "someday" in the mix: "I can't really handle a 40-footer right now, but someday..." "Right now my spouse doesn't like to sail, but someday..." "Right now I only have a week off every summer when I can go sailing, but someday..."

Maybe all those things will change, and your someday will come. So buy the someday boat then, not now. Now, have the boat that suits your present life. If you have time to daysail, own a daysailer. If you have a week off in the summer, don't pay for a big boat all year long— charter. If you really just want to sneak away and cruise the coast, you don't need an offshore boat.

One measure of a boat's worth is how happy it makes its owner, no matter how it's designed or built, what kind of shape it's in, or whether it's underway or not. A lot of people have boats that are too expensive for them, and like them anyway. A lot have boats that have almost infinite work lists and rarely go out—and like them anyway. Those who have little money to buy labor and expert help won't stay around boats long if they can't learn to enjoy tinkering and troubleshooting about as much as they enjoy boating itself. However, a fixer-upper that eventually goes sailing is one thing; a boat fully preened and idle is another. How many times have you seen someone sit at the dock or on a mooring all summer on 10 tons of boat, fiddling with the watermaker, for want of an extra hand? The trick is to have the boat that lets you cast off and go.

How would it be, I wonder, if I were sailing along at six knots on my Ohlson 38 in a crisp fall breeze, with a kettle simmering on the stove and the morning sun glowing through the mainsail, if a Dragonfly trimaran came tearing past at 16 knots in a welter of spray?

Choose Your Seductions

Is one of the purposes of going out on a boat to get away from the hubbub of life on land, or is it to bring hubbub with us? Are we interested in making our lives afloat physically and mentally easier, or do we want to challenge ourselves? Assuming that people's answers to both those questions would fall somewhere along a scale, how much are we all influenced by the marketing people and the engineers who put a feature in a product just because they can? Certainly a lot of products that used to be considered extravagant are now standard; meanwhile boatshows are awash in gear and gadgetry intended to make our boating lives simpler by saving us labor and trouble. So it's important to think about

DISPLACEMENT SAILBOAT PERFORMANCE EQUATIONS

BUILDERS OF DISPLACEMENT SAILBOATS will often publish performance ratio numbers in addition to standard dimensions like length, beam, and draft. These non-dimensional ratios can give you some idea of how stiff or tubby or graceful a boat might be. Two are straightforward: beam-to-length, and ballast-to-displacement. Two others are a little trickier: sail area-to-displacement and displacement-to-length. The first has to do with how much wind horsepower is available to move the mass of boat. The second has to do with how heavy the boat is in relation to its waterline length. Neither of these ratios takes into account the complexities of hull shape, the increase of waterline length when a boat is heeled over, the weight of stores and gear, or a number of other important factors. But they do give decent guidelines for how peppy the essential boat is likely to be.

Here are explanations of the equations, and some guidelines that result:

Sail Area to Displacement:

$$SA / (DSPL/64)^{.66}$$

That's sail area in square feet divided by the result of the displacement (in pounds) divided by 64 (the number of pounds in a cubic foot of seawater), to the two-thirds power. (A scientific calculator is needed to multiply by fractional exponents). Sail area should be calculated with a 100% foretriangle, not including the area from an overlapping headsail.

how much of your labor you want saved, how much of your thinking you want done for you, and how many of your skills you really want to outsource.

How can we resist spending money on the latest electronics? The information we can derive from a late-model GPS chartplotter connected to an integrated instrument system is astounding. Then we can add radar interfaces, weather map overlays, and all sorts of other data, with little icons of our boats superimposed on top. We can couple up a laptop computer with better and

< 16 = very conservatively powered

16-18 = conservatively powered

18-24 = moderately powered

24-26 = high-powered

> 26 = very high-powered

Displacement to Length:

(DSPL/2240) / (.01 x LWL) [3]

That's displacement in pounds divided by the number of pounds in a long ton, then divided by the product of .01 times the waterline length in feet, cubed. Feet/inches have to be converted to feet/decimal feet.

< 100 = very light racing boat

100-150 = light racing boat

150-200 = moderate racer/light cruiser

200-300 = moderate cruiser/racer

> 300 = heavy cruiser

It doesn't pay in the long run to neglect skill in favor of equipment.

---

better cartography and denser data. We can stare at these set-ups for hours and hours—provided we can get it all installed correctly with the wiring right and everything properly calibrated and talking to everything else. And this, too, can take hours and hours. Not long ago, it was a big deal to have your position confirmed by LORAN. It was "all you'll ever need."

It's one thing if you're on a racing boat; it's another if you're on a six-knot family cruiser. I've already spent more than my share of time down below doing

things like getting fluxgate compasses to talk to instruments so that I could have a digital read-out of what I could have seen well enough, if not down to a half-degree, with my own eyes—if I'd been on deck.

Electric winches used to be reserved for folks too rich, weak, or lazy to grind. There's no doubt that if you're sailing shorthanded, or getting older, or have a shoulder injury, these are handy to have. But still, people who are sold electric winches as a matter of course should first have a sense of what it feels like to grind in a headsail by hand. They should know what forces they're dealing with. They should feel the connection between headsail and sheet and winch mechanism and arm and back.

Of course there are no rules. Of course it's a matter of personal preference. But if you're new to boats and you begin your acquaintance by going shopping (a natural move for novices in any sport or pastime), be alert to the Gear Overkill Syndrome and the related sales ploy: Buy More Safety or You May Be Sorry.

Take the example of the electric winch's sister on the bow—the electric windlass, an invention that has put an end to bruising, bleeding and strained back muscles for thousands of sailors and powerboaters. But here's a windlass with a lifting power of over a ton. It draws 110 amps, and costs $3,800. If you're not physically challenged, you don't need such a windlass to pull up a 15-pound anchor and 10 feet of chain.

If you do become convinced somehow that you need a windlass powerful enough to drag your boat bodily off a Tongan reef, that's fine, but with that decision comes a dedicated battery in the bow, or a big, fat copper wire that must be run forth to the bow and back to a big, fat battery bank with lots of amps available, which need to be supplied somehow—by high-output alter-

> A lot of people have boats that are too expensive for them, and like them anyway. A lot have boats that have almost infinite work lists and rarely go out—and like them anyway.

nators or solar panels or wind generators. Complex gear begets more complex gear. Buy what you need, buy what you want, buy what you're comfortable fixing or having fixed. But don't come home the victim of a foisted choice.

Meanwhile there are a lot of products on the market that solve nonexistent problems. For instance, there are various gizmos sold to take the place of knots and hitches. They're used to adjust fender pendants, to cinch and hang coils in lockers, and secure miscellaneous things. They're generally made of shaped or injected-molded plastic, but can be straps with Velcro tabs, or vinyl-covered hooks with pot-metal ratcheting clips, and so on. These things have three strikes against them. First, they can only do the one thing they were designed to do, so they're not versatile, and they take up space when they're not working. Second, they are a needless expense, sold to people who won't bother to learn a basic set of knots, bends, and hitches. It can take a bit of time and effort to learn these things, so some people just sidestep the learning and buy the gizmo. But of the two dozen problems a rolling hitch might solve on board, the spring-loaded Chinese-handcuff tensioner (for 1/4" to 1/2" line—please buy Model B for larger line) will only solve one or two of them—if you happen to have it handy. Third, by resorting to the gizmo, people deny themselves the satisfaction of acquiring a skill that's not only an integral part of the endeavor afloat, but handy in the house, the car, the woods, and plenty of other places. It doesn't pay in the long run to neglect skill in favor of equipment.

Finally there are all the things you can carry out on the water today that

Complex gear begets more complex gear. Buy what you need, buy what you want, buy what you're comfortable fixing or having fixed. But don't come home the victim of a foisted choice.

used to be permanently fastened to the shore, including telephones, TV, and all the wonders of the Internet. The monetary cost of bringing these things on board has dropped sharply in recent years, but what's the cost in hours of lost reading, conversation, observation, skill-building, privacy, peace and quiet? Not long ago, excursions on our boats, whether for weekends or for weeks, immersed us in another world. Once the lines were cast off, we were left alone in a different, fascinating universe with our shipmates and whatever amusements we brought along for the trip. Of course that's still possible, but now there's an almost irresistible pressure to stay constantly attached to the land— to our jobs, to our contacts, to our feeds. We should all resist a bit. And if you're new to boats, give them a chance to help you experience better things than what you can find on a screen.

Boat Show Game Plan

One of the best things you can do to stave off the winter doldrums and start getting psyched for spring is to visit a big boat show. It doesn't matter whether you're in active buying mode or research mode—either way you get that kid-in-a-candy-store feeling, and you'll have fun no matter what. But going to a boat show without a plan is like walking into fiberglass forest. You'll have more fun if you do some prep work ahead of time. Let's work through this.

Chances are, you're going to a boat show for one of five reasons:

- You have a specific boat locked in your sights, and you're going to the show to score a deal.
- You have two or three or four models in mind, and you want to compare them in one visit, with show prices available.
- You're not ready to buy, but you're looking at today's new offerings knowing that there will be used models on the market at reduced prices in a year or two, when you are ready to buy.
- You know you want to get into boating, but you don't really know where to start and just want to see a whole bunch of boats in one place so you can compare them.
- You're shopping for gear or services.

Every boat show has whole reefs of booths with venders selling everything from electronics to teak furniture to sunglasses to insurance and financing.

No matter which of these profiles fits you best, don't just walk into the show without any plan at all—you'll waste valuable time wandering aimlessly in the gelcoat jungle.

Even if you're not in active buying mode, take photos of show prices to help you do your research at home later.

Experienced boat show denizens check out shows online beforehand, so

> Take pictures of the things that are going to concern you if you buy the boat: Can you access the sump pump for the shower in the head compartment? Is the refrigerator big enough? Is there any ventilation in the center-console where you're going to put a toilet?

they know which builders and gearmakers are going to be represented, and sometimes even where they'll be located in the venue. They map out who and what they want to see, figure out a priority list and a walking path, and they come equipped with all or most of the following items:

- Comfortable shoes that can be slipped off and on easily if you're going aboard boats
- A digital camera or cell-phone with a camera
- A notebook and pencil
- Business cards or calling cards to give to sales people or builders if you want them to keep you updated on a boat or product
- A sturdy bag, if you're planning to pick up brochures or take-home boat gear. Yes, show vendors always hand out free bags, but your own easy-to-carry bag, with strong straps and pockets to stow things, will serve better.

There are always, always bargains to be had at boat shows. Builders, dealers, and equipment-makers invest to be at the shows because they know that virtually all the show-goers are truly interested in what they're selling. The question is, what can be done to convert active interest into a sale? The simple answer is discounted prices, but there are other incentives, too, like free or discounted option packages, or attractive financing. Boat show deals can get done quickly because representatives from the OEMs are often there themselves, working with their dealers to make things happen for customers.

Take pictures of the things that are going to concern you if you buy the boat: Can you access the sump pump for the shower in the head compartment? Is the refrigerator big enough? Is there any ventilation in the center-console where you're going to put a toilet? Is there a gasket on the anchor locker hatch? Can you lock the outside stowage areas? Can you get to the batteries easily? Is the fiberglass work in the bilge area smooth and easily cleaned, or has it been left rough and raw—because hardly anyone at a boat show bothers to inspect bilges?

How's the Access to Important Parts?

Since the early days of fiberglass boatbuilding, techniques for making boats strong and leak-resistant—and cheaper and easier to build—have advanced steadily. The accepted method for most boats today is to mold the hull, deck, and possibly an interior pan, as separate units, often with tabs, stringers, bulkhead reinforcements and other details included. The hull is equipped with what it needs, for example an engine, fuel tanks, water tanks, bunks, and so forth, and then the deck is installed permanently on top of it. Today's hull-to-deck joints are fiberglassed to make a monocoque construction, or bonded with extra-tough sealant and stainless fasteners. They are not meant to come apart. Ever.

Yet some boatbuilders, anxious to get boats out the door and make a sale, can be incredibly lazy, or shortsighted, or downright cynical when they seal in equipment that's eventually going to need to be maintained or replaced. When you're looking at new or used boats, be vigilant about these things. Don't let a shiny gelcoat and a good-looking engine keep you from digging a little deeper into your possible future with the boat.

Here are five items that you should be wary of. It's not a full list, but it should serve to get your antennae working.

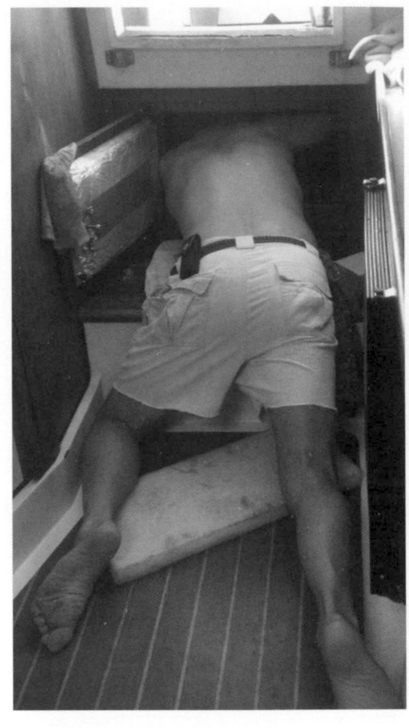

Fuel Tanks - Tanks that are installed under the deck without access points (preferably inspection ports that offer a view straight to the deepest parts of the tanks) are likely to need attention after a certain number of years, especially metal tanks that hold ethanol-laced gasoline. Plastic tanks aren't as susceptible to corrosion, but their connections can still fail. Sometimes the only way to get at them is to saw open the deck above them.

Stuffing Boxes - In inboard boats, changing the packing in a traditional stuffing box is a chore that needs to happen every few seasons, and the packing gland itself will probably need to be tightened even more often. Builders sometimes install items above or around the stuffing box in ways that make it difficult to access.

Plastic Through-Hull Fittings - Plastic through-hull drains at and above the waterline are susceptible to UV degradation. A decade in the hot sun can make them brittle and easily broken. If they fail, you'll end up seeing daylight through a hole in your hull, and that daylight may only be a couple of inches above your waterline. Can you get to every one of those through-hull fittings, or is one or more locked away, out of reach?

Steering Gear - Whether the boat you're looking at has cable or hydraulic steering, figure out how you're going to be able to access the most important parts of it. Can you get to both ends of the hydraulic lines? Can you adjust your cable system easily? Are there chafe points to worry about? Thoughtful builders will install a messenger line in case new hydraulic lines or cables need to be pulled through inaccessible spaces.

Hoseclamps - There's no such thing as a screw-type hoseclamp that doesn't eventually need tightening, whether it's working on seacock plumbing, an engine hose, or a piece of galley equipment. But it's easy to find —- or not find—hoseclamps that have been blocked off behind some other piece of gear or sealed behind a piece of fiberglass, never to be thought of again—until they fail. The more complex the boat you're looking at, the more you'll need to take note of its hoseclamp population.

None of this is meant to scare you away from a used boat that you really think is right for you. Almost every boat is home to a few access problems. Just don't let them surprise you, and if you see a ton of them in one boat, then factor in the time and expense of dealing with each of them if and when the time comes.

Budgeting for Boat Maintenance: How Hands-On Will You Be?

Anyone who thinks about investing in a boat bigger than a kayak should get out a calculator, take a deep breath, and tally up honest estimates of the recurring costs of registration, dockage, hauling and launching, winter storage, insurance, and maintenance. But that last item—maintenance—is a tricky one to figure, because it's not a fixed amount determined by someone else; it will vary season to season and year to year, and it will depend very much on how willing and able you'll be to work on your new investment.

Maintenance, of course, means not only cleaning things, lubricating things, and replacing parts on things that aren't broken, but fixing things that are. And any boat owner who's been in the game for more than a few months will tell you there's a lot of it. There's a strong argument (and I say it's strong because it's mine) that the sense of accomplishment and satisfaction in boat ownership is enhanced by a willingness to embrace the hacksaw. But there's a counter argument that says, "I'm buying this boat to relax on, and I have no intention of spending every weekend busting my transom to keep it the way I want it. I'm planning to pay the pros to do that for me."

Well, good on ya! But let's just make sure you know what you're getting

into, more or less, no matter where you stand along the do-it-yourself spectrum.

Marine labor rates vary widely around the country and the world, as does the quality of work involved, and it shouldn't be surprising that you don't always get what you pay for. You might pay a skilled marine tradesman in the lower Chesapeake or the western Gulf Coast half of what a tradesman or his yard would charge for the same job on Nantucket, Eastern Long Island or San Diego, and you might get a better result to boot. But no matter where you are, you won't find marine labor cheap. Expect a range of $70 to $140 per hour for skilled work (engine work, topsides painting and varnishing, electrical, refrigeration, etc.), and $40 to $80 or so for unskilled projects (pressure-washing, waxing, scraping, etc.). So where you plan to keep your boat will make a big difference in total dollars spent on maintenance, assuming you're planning to pay someone else do so a portion of it.

While boatyard job estimates tend to be low, any good yard manager will make an honest attempt at quoting you a ballpark figure for a job based on

Time + Materials (with a shop markup for materials) or, sometimes, with a fixed price per job, depending on the project and the experience of the yard in handling that job. A yard foreman who's been in the business for 20 years will (or should) have a good idea of how much you'll pay to have your topsides cleaned, compounded, and waxed. That said, boats are complex objects, and even the pros are not immune to occasional task/time-warps that can turn a two-minute project into an afternoon of high-volume cussing. And that time will have to be paid for.

It's obvious that the size of a boat has a bearing on maintenance: The bigger it is, the more expensive. But type matters, too, and so do layers of complexity in terms of systems and surface areas. A 30-foot wooden sailboat with a full keel and auxiliary diesel engine is going to cost more in the long run to maintain than a 30-foot center-console with twin or even triple outboards (although it will probably cost far less initially). The sailboat owner will have spars, rigging, and sails to contend with; an auxiliary engine and running gear; topsides, deck, and bottom that will need regular painting (a full-keel boat takes a lot of bottom paint); plus freshwater, electrical, and sanitation systems to maintain. The expensive center-console, on the other hand, will need to be cleaned and waxed. If it lives in a slip at a dock it will need bottom paint (but less than the sailboat); if it lives on a trailer or on a lift and just splashes around for a few hours at a time, it may not need any bottom paint at all. The outboards will need flushing and periodic maintenance, including oil changes if they're four-strokes. While routine maintenance on modern outboards can be tackled by a moderately handy boatowner, many owners just trailer their boats to their local dealers or service pros, or pay the pros to make house calls for end-of-season servicing. If the boat has a plumbed-through head and holding tank instead of a porta-potty, and a plumbed freshwater supply, those things will take it back up the complexity scale.

So, no matter what type of boat you're shopping for—power or sail; bowrider, express cruiser or center-console; outboard- or inboard-powered;

Boats are complex objects, and even the pros are not immune to occasional task/time-warps that can turn a two-minute project into an afternoon of high-volume cussing. And that time will have to be paid for.

trailer-borne or TraveLift-launched—bear in mind that differences in configuration will make differences in your maintenance budget.

A note on budgeting for materials: There are some maintenance supplies sold at boating stores that can't be found elsewhere, like bottom paint, good-quality marine epoxy, marine-grade electrical connectors, and engine zincs. On the other hand, boat stores make a ton of money with stratospheric markups on items that can be found in regular hardware or home-goods stores— things like mops and hose, paint brushes, high-quality masking tapes, sandpaper, buckets, spray cleaners, and so on. If you're a hands-on person just getting into boats, you'll save yourself thousands of dollars in the long run by knowing what needs to be bought at a boat store and what can be bought elsewhere.

Now we're down to the last element: you, and where you and other members of your family fit on the DIY scale. Probably the easiest way to get down to the nitty gritty is to present a sample list of maintenance chores that might be offered by a full-service boatyard to their clients. Consider the type of boat you're interested in, research the labor rates at the boatyards in the area where you'll keep the boat, picture yourself doing these jobs, then picture trained professionals doing them while you peel fifties off your bankroll.

BOATYARD CHORES: WHO DOES WHAT?

- Fall: Install winter frame and covering. Spring: Remove winter frame and covering.
- Fall: Remove and/or winterize/maintain batteries.
  Spring: Re-commission batteries.
- Fall: Winterize freshwater system.
  Spring: Flush and fill freshwater system.
- Fall: Change engine oil, clean heat exchangers.
- Fall: Winterize engine(s)
  Spring: Flush and commission engine(s).
- Spring: Wash decks; clean windows and hatches.
- Spring: Polish and protect stainless, chrome, bronze, brass, and other metal fittings on deck.
- Spring: Clean, compound, wax topsides.
- Spring: Prepare and paint bottom.
- Spring: Install zincs; prepare props and running gear.
- Spring: Test, troubleshoot, electrical system; anti-corrosion measures
- Fall: Remove and store sails, running rigging.
  Spring: Install sails and running rigging.
- Fall: Remove and store canvas, dodgers, bimini tops;
  Spring: Install canvas.
- Winter: Check bilges, ventilation, boat covers and tie-downs periodically.

Of course, those are strictly recurring maintenance chores; they don't include fix-it jobs, gear replacements, new-equipment installations, or occasional make-overs like topsides painting. At the point where the idea of peeling off the next fifty is just too painful, take a good look at the boat you're lusting after and brace for a decision: You're either going to have to get up off your transom and tackle more maintenance chores than you originally wanted to, or find a boat that will be cheaper to maintain.

CHAPTER 5

The Right Stuff

If you want a pastime or sport that requires comparatively little peripheral equipment, try swimming or running. But if you're a gear nut, have we got a game for you! On boats there's good, proven, must-have gear; there are dozens of shiny objects that cost money and cause distraction while providing little real benefit; and in between are the myriad bits that we scratch our heads over...

Give Me Enough Rope

Left to their own devices, some sailors buy rope the way Imelda Marcos used to buy shoes—impulsively, profligately, with an irrepressible urge. Even today when some of us go to a boatshow we have to stand for a long time next to the booth with the stacked coils of multicolored climbing rope and odds-and-ends in all lengths and diameters, wishing we could come up with a reason to get just a little bit more. There's no such thing as too much. We're melded with Imelda.

In the basement I have everything from spools of whipping twine and tarred marline (which, when you put your nose to it, takes you directly to the fo'c'sl of the Charles W. Morgan) to big coils of nylon anchor rode waiting for a project.

If you have enough rope, projects suggest themselves all the time. Some winters ago, the morning after shoveling off our 57th snowstorm, I got out the gantline I used to use in a four-part tackle to go up the mast. It's a nice soft blue braid, about 160 feet, and I made a rope-tow with it up the little hill in the backyard—put a snap hook in it with a cow hitch so I could pull the lighter kids and their sleds up the hill. The top end went through a block tied to a tree up the hill, the bottom end through a snatch block hooked into a loop around another tree at the bottom. The loop was closed by two rolling hitches, so the whole thing could be adjusted easily. I had more fun than the kids.

Chris Caswell, in a commendable column in *Sailing* a while ago (extolling the habits and skills of seamanship that make sailors so handy—nay, almost godlike—on land) told the story of a friend who needed some furniture moved out of a second-story apartment. Someone retrieved a mainsheet system from a boat in the harbor, and bingo, out the window and down the stuff came.

Rope, rightly rove, can make you look devilishly clever. The more you know about it, and the more you practice with it and rely on it, the more projects and jury-rig solutions will pop up and demand to be tried out. This can actually lead to some hairbrained schemes, like the time someone who shall remain nameless managed to lash a trailer to a hitchless car bumper with something that became, in the space of a mile or so, both Gordian Knot and Fender Bender.

On the other hand, rope, arguably the most versatile tool in the sailor's chest, can and should be used in a lot of places where plastic or metal fittings are now installed in the name of convenience, but at the cost of weight, corrosion, and holes in the boat.

Truly there's nothing so important, so familiar, so comforting to sailors as rope. It's nice to sit down on a winter's night with Clifford Ashley or one of his disciples, and 10 feet of three-strand rope, and work things out. For many, a well-made knot board is a fascinating sculpture, and a Carrick bend is an example of symmetry and strength to rival the most sacred geometry. As Hervey Garrett Smith said in his introduction to *The Arts of the Sailor*, back in '53: "In the final analysis, the pleasures that I have derived from the practice of these skills more than compensate for the endeavor."

In Praise of the Bucket

On small boats, things often need to serve more than one purpose. The bucket is a good example. It can be used as a bailer (nothing more effective at ridding a boat of water than a scared man with a bucket, as they say); as a toilet (fit with an attractive lid); as a stowage bin for wet stuff like masks and fins; as a deck-sluicing tool; as a container for all the little parts you're working with on a mechanical job; as a drogue; as a roll dampening device when weighted and hung overboard from a spar; as a tool container hung from a bosun's chair; as a waste container; as a laundry tub, and for many other things. The trouble is in finding just the right bucket. My wife will tell you that my addiction to buckets is on a par with my addiction to knives, cheap watches, rope, and cool little flashlights. In years past when I was tasked with finding birthday presents for the friends of my children, I've invariably come home with a mini Mag light, or sometimes a hank of colorful parachute cord or 50 feet of flag halyard. "What could be more fun that these things?" I asked my incredulous spouse and incredulous children, who had imagined wrapping up something more along the lines of a Transmogrifying RoboZaur or Mall Lounging Barbie.

I realize that a gray bucket may not strike exactly the proper festive note at

> In impatient hands, a set of locking pliers
> is surely the most destructive device known to man,
> outside the thermonuclear range.
> I keep three sets—big, small, and needle-nose
> (for crushing more delicate objects).

a six-year old girl's birthday party, but really, it should. If the kids only knew.

Once you find a bucket with the right shape (the kind with one flat edge, shaped sort of like a "D", is particularly good on a boat), you have to modify it, usually by removing the metal bail and using the holes to reeve a lanyard. Sometimes you have to drill new holes. If the bucket looks as if it might last for a while, or costs more than $8, you would probably take time to splice the lanyard on. In any case, you need a good bight of line from hole to hole, with a fixed loop in the middle for attaching another line.

There are several close cousins of the bucket on board, related by their ability to take up little space and yet serve lots of purposes. The lowly beach towel can be used as blanket, sun shade, wind block, pillow, sarong, and curtain. The fender can act as marker buoy. It can be trailed behind the boat as a swimming float in a current. It can be thrown to a man overboard. The boathook already has a dozen well-known uses: Aside from grabbing mooring pendants, it pulls things in, fends things off, wings sails out, supports awnings and sun shades (see beach towel, above), grabs dancing errant halyards, and now, in the guise of the "multi-purpose boat pole" it swabs, squeegees, and paddles. The point is, space aboard a boat is always at a premium, and the more purposes a bit of gear can serve, the more you come to appreciate it. It's unfortunate that the redundancy and clutter we strive so mightily to eliminate aboard boats is regarded so differently ashore.

The other day, friends came over to repo the sofa they'd left with us for a couple of years. Now there's a big open space in our living room, and my wife is thinking of a way to fill it. I keep telling her that we already have a chair, and if she needs it where the sofa used to be, I'll be glad to move it over for a while if someone wants to sit there. Better yet, we could get three or four buckets and put boat cushions on top. That way we'd have… more buckets! Also spare throwable devices. Or maybe three buckets, up-ended, with folded beach towels on top. Sure, that's the ticket.

Tools

Tools, obviously, are of critical importance to most sailors—at least to sailors who wreck their own boats rather than pay someone else to do it. (Around the house my wife calls me Measure Later Logan, but she fails to see how a fix-it attack that is both impatient and bullheaded can lead to greater improvisational skills later on.)

For many of us, tools, like rope, are things we can't get enough of. When you mangle things on boats yourself, you need a good selection of harmful tools. In impatient hands, for example, a set of locking pliers (ViseGrips) is surely the most destructive device known to man, outside the thermonuclear range. I keep three sets—big, small, and needle-nose (for crushing more delicate objects).

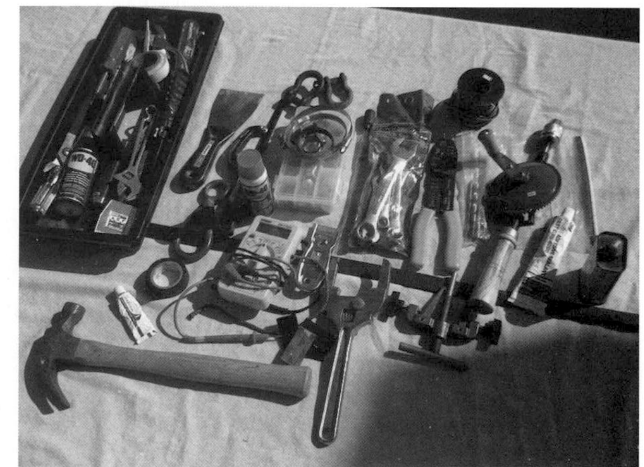

A good rule of thumb is that a sailor's toolbox need not exceed his capabilities. Even then it helps to follow the principle of Toolkit Targeting, which says, in a nutshell, "Don't ship the shop. Know what you'll need, and just ship that."

A sailor's toolbox is a sacred thing, and I would not presume

RECOMMENDED TOOLS FOR DIFFERENT-SIZED BOATS

For dinghies and very small boats near right along shore:

- A corrosion-resistant multitool, e.g. Leatherman, Swisstool, etc.

For smaller sail and powerboats with a bit of stowage space, add:

- Four-way screwdriver (large and small Philips and slotted)
- Adjustable wrench
- Lubricant (WD40, Liquid Wrench, Teflon spray)
- Duct tape

For larger near-shore boats with more stowage space, add:

- End wrenches, socket wrenches, Allen wrenches (hex keys) in the sizes found on the boat. Not necessary to store other sizes.
- Assorted pliers: regular, needlenose, channel-lock, and locking (ViseGrip)
- Wire cutters/snips
- Claw hammer
- Wooden or rubber mallet
- Flat and round files
- Hacksaw with extra blades

to say absolutely what should or shouldn't be in there. I can only say which tools have worked well for me, and these are listed in the accompanying box. Beyond those basics it's possible to expand almost indefinitely, depending on the mission of the boat, the skills of the crew, and available space. Many sailors would go for a larger power-tool collection: angle-grinders, reciprocating saws, Dremel tools, and sanders are common.

Another handy tool is the two-foot, flexible, spring-bodied, plunger-activated grabber-claw gizmo that you use to fish wires and messenger lines through things, or retrieve screws you've dropped in inaccessible places like a

- Small wood saw
- Razor blades
- Dacron cord
- Tape measure
- Electrical tape
- Small hand drill and assorted bits
- Wire stripper/crimper
- Multimeter
- Hand clamps and screw clamps
- Hose clamps
- Wooden bungs for through-hulls

For offshore and long-distance boats, add:

- Cordless drill/driver
- Inverter (300-watt or higher) for charging battery-powered tools
- A bench vise, if there's room
- Propane torch
- Engine spares (fuel and oil filters, water pump impellers, gaskets, assorted hose, belts)
- Whipping twine; sailmaker needles, wax, and palm

crevice in the bilge. You can often do as well with a dowel or long screwdriver with a bit of duct tape fastened on the end with the sticky side out.

On bigger boats there will always be a self-expanding collection of tapes, lubricants, sealants, and fasteners. Try to keep them separate—a bag or box for goops, another for tapes and Velcro, another for fasteners and small hardware. Another box, bag, or bucket can hold cleaning gear—brushes, sponges, rags, cleaners, waxes, rubbing compound.

A word on sandpaper. Unless you absolutely have to keep on board traditional aluminum oxide sandpaper for painting or varnishing projects (or silicon carbide for metal), the wet/dry sanding sponges and pads made by 3M will work on the great majority of projects. They don't curl up or rot, and are reusable many times.

Finally, there should be a sharp knife on deck, easily reachable by any crewmember. On sailboats it can be sheathed at the mast; on powerboats somewhere in the cockpit.

I had an old pressurized alcohol stove that seemed to work most enthusiastically when it was spreading colorful sheets of blue flame across the cabin.

Not Your Mom's Kitchen

"Tell me what you eat and I shall tell you what you are." That was the gastronome Brillat-Savarin, and his phrase eventually morphed into "You are what you eat." If this is true, then I am an amalgam of canned hashes steeped in Heinz ketchup and Kentucky sour mash. Still, I claim some refinement of taste, at least on shore, where I wouldn't go near canned peas if it were the last food in the world. On the water, though, my standards sink to survival level. I can remember many nights after long, cold days of racing when we hunkered down shoulder to shoulder in the cabin of that Tartan 27 with dry socks on, the lamp going on the bulkhead, the rain pattering quietly overhead— and those Le Sueur peas were ambrosia of the gods.

I also have a vivid picture in mind of a meal in one Bermuda Race when we were completely becalmed and sweltering, south of the Gulf Stream, after working our way through a long gale. The off-watch sat below, a bit stultified, looking at our heroic sea-cook stirring a big pot of hot lunch. He had a half-burnt cigarette dangling from his mouth and a bead of sweat rolling back and forth across the tip of his nose at the same frequency as the glassy swell that made the boom slat with a quick and grating screech every few seconds, no matter how hard we prevented it. We all watched hungrily, wondering whether the sweat or the ash would drop first into the burgoo. Eventually they went in at about the same time, and we all dug in like trenchermen.

When I lived aboard, I had an old pressurized alcohol stove that seemed to work most enthusiastically when it was spreading colorful sheets of blue flame across the cabin. I spent far more time cleaning its crumbly burner assemblies than cooking on it. I would set the thing going when I got home from work, and put on a saucepan of water. Into the water I'd put a can of Chef Boyardee spaghetti with a couple of vent holes punched in the top. An hour or so later the water and spaghetti would be lukewarm. I'd take the top off the can, eat the contents with a fork, toss out the can, wash my face and the fork in the lukewarm water, pour the water down the head in hopes of unfreezing it a bit, and then climb into my sleeping bag for the night with my eyes watering from the fumes of unburnt alcohol. The contented privations of youth.

Eating aboard is easy; stocking and managing the galley, cooking the food, and washing up—especially underway—is not. Anyone who manages a galley well is a shipmate to be deeply treasured. I've known a few—people who could make a meal for eight while heeled over 20 degrees and pounding through waves. People who regularly baked bread on board. People who could use a gallon and a half of fresh water to wash settings for six, and the pots and pans as well, and properly.

A sea-cook has an unerring sense of which way a hot liquid will fall in a tilted world full of roll, pitch, and yaw. Some people just can't manage it, and must be given a wide berth. Aside from the fact that most galleys these days are not well suited to cooking at sea, few people are willing to put up with the challenges anyway. Dealing with pots and pans that are always nested; discovering by trial and error just the right combination of tools and utensils to have

GALLEY GEAR

Not long ago we were rafted up with two other boats and it was our turn to cook breakfast for six people. We made soft-boiled eggs and handed them out in paper cups, along with a cutting board with toast, butter, marmalade, and salt and pepper. Everyone peeled their own eggs, kept the shells for their own gurry buckets, and ate the eggs out of the cups. We then stacked the empty cups and put them in our trash. This solution was better than washing six plates with hardened egg on them, especially with our limited water supplies. We also had clean hot water in the egg pot to do our silverware and bread knives. In general, though, if you have plenty of water available it's more civilized and easier on the environment to use washable dinnerware made out of heavy-duty plastic.

You can easily overstuff a galley with pots, pans, and dishes that rarely if ever get used. If you cruise with four people aboard and maybe have one or two more occasionally, you don't need place settings for eight. In fact, when we cruise with friends and are invited aboard each other's boats for drinks or meals, people often move from boat to boat with their own eating and drinking gear. Try starting with just the following. If it's not enough, add more.

- Each crewmember gets a mug, a plate, a bowl, a fork, a knife, and a spoon.
- Napkins all around; paper towels work

aboard, and how they need to be stowed; working on minimal counter space; rationing cooking fuel, fresh water, and the food itself; constantly organizing and choreographing moves so that everyone eats hot food on time, with no one burned or stabbed—these are the talents of a good sea-cook.

On the Edge

Knives on boats are hugely important, and I'm sometimes amazed to discover that I'm the only person aboard carrying one. Actually, it's a Victorinox Spirit

- Stove fuel, with a spare supply
- Two cooking pots with lids; small pot stacks in large pot
- Frying pan with lid
- Three nested plastic mixing bowls, large, medium, small
- Cutting board
- Dish rack
- Cloth dish towels and paper towels
- Large carving knife with sheath
- Small serrated knife with sheath
- Two serving spoons, one slotted
- Spatula
- Baster (can also use to top up battery cells)
- Can opener
- Vegetable peeler
- Flat sieve
- Kettle
- Coffee-making system of preference—drip, percolator, or French press
- Thermos
- Moisture-proof containers for salt, pepper, sugar, etc.

multi-tool in a belt sheath, and I literally go nowhere without it, except to bed and the shower, and even then I call to it once in a while. I hate having to check it in my backpack when flying, because it seems to me that hurtling through the air in a complex mechanical device, with the thought of suicidal maniacs aboard, is exactly the situation in which a multi-tool might provide a measure of comfort. (Talk about Walter Mitty.)

In any case, even a multi-tool is a compromise on board when you're under sail and the running rigging is loaded up. In exchange for the versatility of the multi-tool, you give up a degree of whip-it-out safety. Same with carrying a standard folding rigging knife or Swiss Army knife on a lanyard in your pocket. In a breeze, whether passagemaking offshore or in a race around the buoys, it's good to have a fixed sheath knife on your belt, situated near the middle of your back so that you can reach the handle with either hand. It's an old picture, but it ain't hokey.

If a person has a hand sucked into a heavily loaded halyard winch, or there's an override on your primary winch and the spinnaker sheet winch is occupied, with minimal steering and a collision imminent; or you're suddenly dragging for fish with your spinnaker and you find that someone has tied knots in the ends of sheet and guy; or you're on your beam ends with the boom dragging through the water and beginning to look odd, and you can't release the preventer, and the bloody thing refuses to break... well, these are all times when a sheath knife needs to be deployed smartly and put to work. The couple of times I've seen things like that happen, the decision to cut has taken about three seconds, and the grabbing of the knife and the cutting—maybe another three. Six seconds can be a long time in an emergency, especially if there's a limb involved. See how long it takes to get your blade out and ready to cut. If you're comfortable with the time, good.

Aside from everyone's personal knives, there should be a couple of others strategically positioned around the boat. Some big-boat sailors keep one strapped to the solid vang, or in a winch handle holder at the mast partners. Some people strap them to the pedestal. If you do that, put it down low, and check for magnetic influence. A cubby locker in the cockpit coaming is a good place. I keep a folding rescue knife Velcro-ed right in the companionway.

The traditional sailor's blade shape has been a sheepsfoot, in which the top of the blade curves down to a point at the front of the bottom edge. This is partly to avoid having a dagger point out front on a lurching deck, but also because the top of the blade is dull, and can be handled, and the blade surface is big enough to let the knife act as prybar, stirrer, and eating utensil. Some sailors, however, think the drop-point blade is just as versatile in other ways, and are willing to give up a degree of safety for it. Do settle on some kind of personal knife or multi-tool, though, if you don't normally carry one. It's one of the most important pieces of safety equipment you can have on board, or anywhere else, for that matter.

Lights and Shades

Aside from a multi-tool or a knife, there are three other crucial on-deck personal effects that every sailor ought to have—a watch, sunglasses, and a pocket flashlight or bite-light. The watch, like the multi-tool, stays on all the time. Luckily, the sunglasses and flashlight can be carried at opposite ends of the day.

It's no good having one or two big flashlights aboard that everyone can share—along with those, everyone ought to have his or her own, and be able to put their hands on it immediately. Today's pocket-sized LED flashlights are great. Some are waterproof, some have strobe or SOS functions, narrow or wide beams, lanyards, clips, and red-light options. Get one that runs on AA or AAA batteries, which are a lot easier to find than button cells.

As for sunglasses, if there's any population in the world who needs them, and good ones, it's the boating population. Direct and reflected sunlight can

bered how shocked Barber had been when Edwards told him he didn't know what kinds of trees grew in his yard. This lack of fundamental knowledge was appalling to Barber. How could someone not know the names of the trees that grew around him? As Barber felt about trees, I feel about direction. How can anyone, particularly a sailor, be complacent enough to neglect the law of ancient, unfailing magnetism, and the instrument that gives it order?

Paper charts are no anachronism, either, when it comes to having a full quiver of navigational arrows. While NOAA no longer prints them, big flat charts can still be ordered from private suppliers online, complete with the most recent NOAA updates. Aside from being things of beauty, they give a much more inclusive field of view at virtually any scale compared to what you can see on a plotter or computer screen. Even in book or big spiral-bound format, paper charts display data immediately, in color, and at the proper resolution. You can draw on them, make notes on them, and transport them easily.

> When a whiff of spruce in the fog corroborates your plot and helps connect you to all the invisible things around, then dead-reckoning will be even more of a thrill than that gorgeous new chartplotter glowing in your helm station.

They require no electricity, except maybe a light for night reading.

They also do a better job at orienting you when you're in a far-flung locale without local knowledge. (The Virgin Islands, the Bahamas, the Adriatic, the Sea of Cortez, the Pacific Northwest, the Great Lakes, or New England—any place distant enough from home waters to require you to study a chart is flung far.) If you've never been to a place, you can't orient yourself until you get your bearings, literally. Even then, it can be difficult to relate what you're seeing to what's on the chart. Sometimes hillsides, harbor inlets, breakwaters, buoys standing offshore, and other things that are clearly marked in two dimensions just don't look as familiar in three dimensions.

In days of old, surveyors went to a great deal of trouble to sketch the profiles of islands and coastlines they were exploring. James McNeill Whistler, for example, before he painted his mother, was a US Navy cartographer, and some of the profile etchings we see today on reproductions of old charts are his. I imagine that some mariners of old, when they were given charts with these beautiful, detailed profiles in the margins, may have griped, momentarily, the way some of us do now: "Drawings, eh? Why, back in my day we made sure of our bearings and soundings, that's all. Never had any of this new-fangled picture art to help us along. Next they'll be bringing the islands to us. Might as well stay home!"

SPEED, TIME, DISTANCE

A FEW YEARS AGO I was driving along I-95 just west of New Haven (on a stretch once described to me by a long-haul trucker as "the worst piece of road on the continent, including Baja.") in moderately thick traffic that was moving along at about 60 miles an hour. I was in the right lane, with a lot of space between me and the car in front. A huge SUV came charging down an open space behind me at about 80, swerved into the left-hand lane, hit the brakes, and then camped on the rear bumper of the last car in line there. In the course of the next several minutes, this guy changed lanes many times, was never more than about four feet away from the car in front of it, and by dint of some breathtakingly risky maneuvers, got about a quarter-mile ahead of me before getting off the highway. I took the same exit—about 15 seconds later, and ended up behind him at the next stoplight.

I'm sure I ended up just as stressed as whoever was in the SUV, because I have a severe adverse reaction to people who have no clue about the relationship of speed, time, and distance. Sailors know the equations:

Time x Speed = Distance.
Distance / Time = Speed.
Distance / Speed = Time.
Convert to and from decimals as needed.

In a watery world with no road signs and few distinguishing marks, these equations, or at least the import of them, become deeply ingrained, such that one eventually becomes aware of the relationships in all situations, no matter what vehicle or medium one is traveling in. Good racing sailors have

Today, one very good reason to own paper charts in book or spiral-bound form is that they often come with aerial photos of harbors and passages.

Just to slug the dear departed horse one last time, no one in his right mind would forswear GPS in favor of paper charts, but dead-reckoning skills are a well-tuned sense of these things—they can judge the distance to a starting line or the time to a layline with remarkable accuracy. Practiced long-distance navigators are similarly accurate, whether they're Polynesians using the movement of stars and the cadence of wave sets, or Joshua Slocum using a windup clock, or Horatio Nelson using a timer and logline.

The original Dutchman's log was just that—a piece of wood that the navigator tossed off the bow. He walked aft, counting the seconds until it reached the stern, then divided distance by time to learn the speed of the vessel. For example, say a 60-foot vessel takes 5 seconds to sail past the log. That's 12 feet per second, 720 feet per minute, 43,200 feet per hour, divided by 6,076 (the number of feet in a nautical mile)—or 7.1 knots. Of course, a smart navigator on that 60-footer would develop a simple reference table.

In 1637, English navigator Richard Norwood figured out a system composed of a line knotted every 47.25 feet and a 28-second sand timer: If the first knot in the line passed through the mate's hand just as the sand ran out, the ship was going 6,076 feet per hour, or one knot. After that, presumably, the oceans became less littered with wood chips.

We need to keep manual navigation skills tuned up, even as we welcome cool developments in electro-navigation. It's not just about safety, simplicity, expense, electrical demands, water-resistance, and all that. Some ingrained practices lead to better awareness of surroundings, of what's possible and impossible, practical and impractical, maybe even wise and unwise. If we can just stay familiar with the relationship of speed, time, and distance, maybe we won't end up like the driver of that SUV—stupider than a moth flinging itself against a lightbulb.

valuable in dozens of other ways. They aren't just boat skills; they're life skills. Knowing them and practicing them once in a while on a paper chart with a compass, parallel rules, dividers, and some simple math gives you a more intimate understanding of what's going on around you than a chartplotter can.

> There was a dark-red, striated smear
> looming below the ridge to our east.
> The smear turned crimson, then pink.
> The sun appeared briefly, an intense yellowish-rose,
> then disappeared completely as the day began.

I'm not talking about the quantity of data or even the near-perfect accuracy of data; I'm talking about taking a hand in producing the most important data yourself, with your own observations of your surroundings, your course, and your speed. And when a whiff of spruce in the fog corroborates your plot and helps connect you to all the invisible things around, then dead-reckoning will be even more of a thrill than that gorgeous new chartplotter glowing in your helm station.

As the World Turns

The tides, given to us by a continuous rolling squeeze of the moon, and in many instances the sun, have always been fascinating to scientists and coastal sailors. Where I live, we have semi-diurnal tides, spaced about 6 hours and 12 minutes apart. Simple. But it's also Long Island Sound, where water rushes in and out at the two ends—Hell Gate to the west and Plum Gut and The Race to the east—in ways that create big local variations all up and down the Sound. Add river mouths, depth variations, persistent winds, and it gets more interesting. Depending on your location on the planet, there are also diurnal moon tides—meaning one high and one low per day, and mixed tides of alternating diurnal and semi-diurnal—and even, apparently, strictly solar tides, as in Tahiti.

As I write this, it's near the winter solstice. A few minutes ago there was a dark-red, striated smear looming below the ridge to our east. The smear turned crimson, then pink. The sun appeared briefly, an intense yellowish-rose, then disappeared completely as the day began, and will stay, in the classic bone-chilling shade of gray that seems to be the work suit of southern New England in December. Tonight we'll have some freezing rain.

It's remarkable that we have our own star so handy. Even as the ridge rolls right to left just underneath its glow, the star seems to move in a thumbnail arc on its way to settling this afternoon behind the woodpile. It won't bring us much warmth, because our end of the world is tilted away from it. But in a few days, our path around that star will start bringing our end toward it again, and we'll be getting warmer. For now our friends Down Under are reaping its rewards.

We're privileged, in sailing, to have working relationships with the earth, moon, and sun—and, for some celestial navigators, with the planets and other, more distant stars.

What I noticed was that wonderful dynamic
of the wind fluid and the water fluid and
the responsive boat moving between them.
I was delighted just to be in the middle of it,
almost like an observer, although somehow
my hand played a part in it.

CHAPTER 7

The Tao, The Hsü, Whatever

A lot of what we do on the water can be pretty goal-oriented, whether it's racing for victory, or passagemaking to a distant harbor, or fishing for the big catch. And the pursuit of goals of course allows you to see the sights along the way. But there's something underneath everything else when you're on a boat doing what you love. It's an obvious thing, and yet it's tough to find words for it. In sailing maybe it's that feeling of traveling by means of your own skill, right along the junction of air and water, using a device as simple as a sailboard or complex as a tall ship, to dig down a bit into the liquid for leverage, and up into the air for power, and go where you want.

I rediscovered that mysterious feeling after one racing season ended, when I was about to put the Laser away for the winter. It was a warm fall day with a light southeast breeze and a bit of haze. I just went for a sail, with no one to compete against, and nowhere I had to go. I went in a big loop, for miles. There was an occasional puff that let me hike out, but most of the time I could just sit with tiller in one hand and sheet in the other. Somewhere in the middle of this sail I forgot about the job, the bills, the insurance, the news of the day, the rush of demands, and just sailed on. The boat didn't have to be trimmed perfectly and I didn't have to think furiously about my next move. What I noticed, but without any particular sense of urgency, was that wonderful dynamic of the wind fluid and the water fluid and the responsive boat moving between them. I was delighted just to be in the middle of it, almost like an observer, although somehow my hand played a part in it. It lasted an hour or so, and was worth six months of the grind.

That contentment, or "oneness," I think, underlies everything in this calling to boats. We all must have felt it from time to time, or we wouldn't be involved.

Ship-to-Shore

If you're a fan of Patrick O'Brian's books, you'll remember some of the scenes throughout the Aubrey-Maturin series when Jack Aubrey is ashore and his house and grounds are being tended by some of his ship's crew as they all wait to get afloat again. I love those scenes, because they remind me of why the water is better than the land. Aubrey's tars, used to rigorous schedules and maintenance chores at sea, are a bit like fish out of water as they attempt to straighten and square away every stone and doorjamb, cover everything in paint and whitewash, and generally make the somewhat disheveled estate conform to the geometry of a warship in the Royal Navy.

These attempts are funny and endearing, at least to me. I like the ingrained habits and disciplines of my fellow sailors, even if they can seem a bit compulsive, if not hidebound, to those who don't have to worry about whether they'll waft down to Davy Jones' locker, drowned, by the end of the day, because of something carelessly left undone or askew. My friend Kim, whose family sailed for many years with Rod Stephens (that godlike figure of Corinthian racing and twentieth-century yacht design), says that the highest praise Stephens could bestow on a piece of gear or an onboard system was, "I don't see how that could be improved upon." It was a rare commendation, and an indication of the sailor's mentality—always looking for ways to make something smoother,

This pastime, this calling, is a microcosm of many of the most fascinating things in life. It encompasses natural wonders, both constant and occasional, and adventure, both quiet and hair-raising. It gives us an honest and encouraging view of our place in the universe.

tighter, hardier, safer, smarter. Kim has a habit, to this day, of modifying much of what he buys—because most everything can be improved upon, at least for his purposes, if only by the addition of a lanyard hole, an extra dart of fabric, or a reshaped handle.

Seagoing habits of organization serve very well ashore, as do habits of observation, calculation, self-reliance, and curiosity. The habit of learning new things is also pronounced, even if unintentional. Intensive learning can't be helped if you spend any appreciable time near other sailors and the sea. Eventually, we learn a bit about oceanography, meteorology, and marine biology. We learn something of electricity and mechanics. If we travel far enough, we learn about other nations and peoples. We learn first-hand about the foulness of pollution and hyper-production in a throwaway culture, and the importance of renewable energy and conservation. The habit of paying attention to these things is naturally carried from ship to shore, where it makes life on terra firma more interesting.

This pastime, this calling, is a microcosm of many of the most fascinating things in life. It encompasses natural wonders, both constant and occasional, and adventure, both quiet and hair-raising. It gives us an honest and encouraging view of our place in the universe. (When you're a speck, you get perspective.) It then connects us with the history of our predecessors as they harnessed the Cartesian view of the world, and the Copernican and Galilean views, to practical methods for finding their way around the planet.

In the close confines of a boat we learn more in a day about the people we're with than we would in a month on land—or maybe ever. Friendships forged afloat are the strongest. But it's those little routines and disciplines, including the urge to improve upon, that make all the difference, day to day, even if they sometimes seem out of place ashore.

I suspect that the sailor's fondness for simple, geometric arrangements, and the predictability of routine, and the compulsion to square away everything within reach, is in fact an acknowledgment that chaos is lurking just around the corner in the form of a line squall, or maybe a visit from a mother-in-law or tax auditor, and he will at least take comfort in knowing where to lay his hands on things, and where to step, when the darkness descends and rough stuff starts.

CHECKLIST FOR LEAVING THE BOAT

__ Center rudder and lock helm.

__ Check sail cover and its fasteners.

__ Coil and stow all lines and running rigging.

__ Double-check security of mooring/ dock lines.

__ Check chafing gear.

__ Check for open ports, especially in cockpit well and hull sides.

__ Ensure adequate ventilation through ventilators.

__ Frap external halyards away from mast with gilguys to the shrouds.

__ Stow ensign and any loose gear.

__ Check seacocks and close if necessary.

__ Shut off batteries.

__ Check padlocks on lockers and companionway.

__ Make sure any shore power connections are secure.

__ If at the dock, manhandle the boat to make sure bow, stern, and spring lines are set to proper length and angles.

Many people will cleat their mooring lines, then spend a while contemplating things, maybe tweaking a line or two, then walk away, hoping that they've set things up right.

The better move is to lay hold of that gunwale and haul away, forward, aft, and athwartships at bow and stern. Try hard to make the boat hit something, because that's what wind and tide will try to do. Then you'll know if your lines are set up right. And it's good entertainment for the neighbors.

About the Author

Doug Logan has been managing editor, technical editor, and executive editor of *Sailing World*, webmaster for *Cruising World*, contributing editor of P*ower-boat Reports*, editor-in-chief of *Practical Sailor*, and senior editor at the websites of the Boat Group. He has written hundreds of articles and edited of dozens of books about boats, sailing, and the sea. He lives in Stony Creek, Connecticut.

A Few Good Reference Books

Chapman Seamanship and Piloting by Elbert Maloney

The Annapolis Book of Seamanship by John Rousmaniere

Boatowner's Mechanical and Electrical Manual by Nigel Calder

Manual of Offshore Cruising by Jim Howard

The Practical Encyclopedia of Boating by John Vigor

Offshore Sailing by Seifert and Spurr

The Seaworthy Offshore Boat by John Vigor

Desirable and Undesirable Characteristics of Offshore Yachts, John Rousmaniere (editor)

Sailboat Electrics Simplified by Don Casey

How Boat Things Work by Charlie Wing

Powerboater's Guide to Electrical Systems by Ed Sherman